Optical and Laser Diagnostics

Other titles in the series

The Institute of Physics Conference Series regularly features papers presented at important conferences and symposia highlighting new developments in physics and related fields. Previous publications include:

Optical and Laser Diagnostics

First International Conference on Optical and Laser Diagnostics held in
London, UK, 16–20 December 2002

Edited by
C Arcoumanis and KTV Grattan

Institute of Physics Conference Series Number 177

Institute of Physics Publishing
Bristol and Philadelphia

British Library Cataloguing in Publication Data

A catalogue record for this book is available from the British Library

ISBN 0 7503 0958 X

Library of Congress Cataloging-in-Publication Data are available

Publishing by Institute of Physics Publishing, wholly owned by the Institute of Physics, London

Institute of Physics Publishing, Dirac House, Temple Back, Bristol BS1 6BE, UK
US Office: Institute of Physics Publishing, The Public Ledger Building, Suite 929, 150 South Independence Mall West, Philadelphia, PA 19106, USA

Printed in the UK by MPG, Bodmin

Preface

Optoelectronics and, in particular, laser diagnostics are becoming an integral part of engineering and medical instrumentation presently under development or already in use in the industrialized world. From the automotive industry to the monitoring of blood flow in veins and arteries, optical techniques are used in increasing complexity and at reduced cost to provide information about the operation of systems considered until recently as "black boxes".

This Conference, the first organized at City University by the School of Engineering and Mathematical Sciences on the subject of Optical and Laser Diagnostics, is bringing together in-house expertise on optical sensors and industrial laser diagnostics and aims to become an international forum for the presentation of ideas on the development and application of laser techniques. These tools are considered necessary to maintain "cutting edge" research both within and across traditional engineering and scientific boundaries, but are also essential in the refinement of diagnostic techniques for condition monitoring, model validation and system optimization.

The Conference is organized into several sessions, each starting with a relevant invited presentation by an expert in the field and containing a number of contributed topical research papers. The material presented covers important themes such as laser diagnostics, velocity measurement, and engine-related applications, and also includes consideration of both biomedical applications and a wider spectrum of topics in optical and laser diagnostics. The papers draw their authority from the expertise of their authors, coming as they do from a number of major universities and industries in the UK and beyond, where such research work is carried out.

The support of the Conference from the industrial sponsors, and the hard work of both the local Organizing Committee and the contributions of the International Committee are greatly appreciated. In addition, the Institute of Physics, and in particular Claire Pantlin of the Meetings and Conferences Department, are to be thanked for their support for ICOLAD 2002.

C Arcoumanis and K T V Grattan

City University, London

Contents

Section 3: Laser and Optical Diagnostics

Section 4: Velocity Measurements

Section 5: Biomedical Applications

Inst. Phys. Conf. Ser. No. 177
Paper presented at 1st Int. Conf. on Optical & Laser Diagnostics, London, 16–20 Dec. 2002

Advanced Laser Spectroscopy in Combustion, Catalysis and Medicine

Jürgen Wolfrum

Physikalisch-Chemisches Institut
Universität Heidelberg
Im Neuenheimer Feld 253
69120 Heidelberg

Abstract. In recent years, a large number of linear and nonlinear laser-based diagnostic techniques for nonintrusive measurements of temperatures and species concentrations with high temporal and spatial resolution were developed and have become extremely valuable tools to study many aspects of combustion processes. Data gained from such experiments are the basis for comparison with detailed mathematical modelling of laminar and turbulent reactive flows to find optimal conditions to lower pollutant formation and fuel consumption. As examples the formation of nitric oxide and ignition processes in automobile engines as well as NO-reduction in municipal waste incinerators are described.

Nonlinear laser spectroscopic techniques using in-situ infrared-visible sum-frequency (SFG) generation can now bridge the pressure and materials gap to provide kinetic data for catalytic reactions. This is illustrated for reactions of CO and NO on single crystal and polycrystalline platinum surfaces in a stagnation flow arrangement.

Finally experiments on the kinetics of enzyme reactions using single DNA molecules in microcapillaries are presented. Single dye labelled nucleotide molecules can be distinguished after exconuclease cleavage by recognition of the fluorescence lifetimes using a confocal microscope and pulsed semiconductor lasers. This allows the construction of hairpin nucleotides and peptides for the early and ultrasensitive detection of infectional and tumor diseases.

Inst. Phys. Conf. Ser. No. 177
Paper presented at 1st Int. Conf. on Optical & Laser Diagnostics, London, 16–20 Dec. 2002
©2003 IOP Publishing Ltd

Local Flame/Flow Characterisation in Turbulent Flame Propagation

S Jarvis, G K Hargrave

Wolfson School of Mechanical and Manufacturing Engineering
Loughborough University
Loughborough
Leicestershire
LE11 3TU
UK

Abstract Experimental laser diagnostic data is presented for flame characterisation during interactions with turbulent fields generated in the wake of a solid obstacle High-Speed Laser-Sheet Flow Visualisation and Twin-Camera Digital Particle Image Velocimetry were employed on a propagating flame interacting with a singe rectangular obstacle. A technique for calculating flame displacement speed was developed and gave results as high as ten time unstretched laminar flame speed. Negative values in agreement with other researchers were also measured

1 Introduction

The quantification of the role of turbulence as a mechanism for increasing the speed of propagation of a premised flame has been of continuing interest for both academic and industrial research. The need for quality experimental data to increase the understanding of such flames and help validate mathematical models is recognised.

It is clear that by fragmenting the flame front and increasing the flame surface area, turbulent flow structures in a propagating flame can increase the local flame speed and hence burning rate. It is well understood that an important mechanism for the generation of such turbulent flow structures in explosions is the interaction of a propagating flame with a solid obstacle. Moen et al (1980) showed the influence of downstream flow structures generated by orifice type obstacles on increasing flame speed. Recent advances in optical diagnostic measurement techniques have provided further understanding of premixed turbulent combustion. Studies by Fairweather et al (1996) provided data in small scale experiments aimed at investigating the interaction between propagating flames and solid obstacles. Linsdtedt and Saktitharan (1998) presented two component velocity data on the interaction of premixed flame and wall type obstacle.

Hargrave et al (2000, 2001) investigated the effect of mixture stoichiometry and obstacle geometry on the flame structure during interaction with a single centrally mounted obstacle. The work showed clearly that the acceleration of a propagating flame was enhanced by its interaction with the solid obstacle.

In order to quantify the interaction between propagating flame and turbulent structures many past researchers have detailed the use of Flame Displacement Speed (S_d) when calculating the effect of stretch on a flame. Sinibaldi et al (1998) presented the result:

$$S_d = V_f \cdot \overline{n} - V_r \cdot \overline{n}$$

where $\overline{n}$ is the unit normal to the flame segment, V_f is the velocity of the flame and V_r is the velocity of the reactants ahead of the flame. They experimentally measured flame displacement speeds from 0.7 to 5.25 times the unstretched values.

The aim of this paper is to investigate propagating premixed flames and provide data, and comparison data for displacement speeds during the interaction of flames with turbulence generated in the wake of solid rectangular obstacles.

2 Experimental

2.1 Experimental Rig

In the present study a square cross-section combustion chamber was employed. This was 150 mm by 150 mm in cross section by 500 mm long and incorporated a single rectangular obstacle 75 mm by 150 mm by 10 mm, located at 100 mm above the base plate. The obstacle presented a 50% area blockage. The chamber was closed at the lower ignition end and open at the opposite end. A thin plastic membrane. which failed shortly after ignition, was placed across the open end to contain the premixture. Hargrave et al (2002) detailed the flow delivery system and mixture preparation of the stoichiometric premixed Methane/Air investigated.

2.2 Laser Diagnostics

For the present study two optical diagnostic techniques were employed. High-Speed Laser-Sheet Flow Visualisation (HSLSFV) provided a global visualisation of the flame propagation and interaction. Twin-Camera Digital Particle Image Velocimetry (TCDPIV) allowed the detailed characterisation of the instantaneous flow field generated.

The HSLSFV setup, detailed by Hargrave et al (2002) incorporated a Copper Vapour laser synchronised to a Kodak 4540 High Speed Motion Analyser. The recording system was triggered by the mixture ignition and provided imaging at 9000 Hz. Olive oil seeding introduced into the premixture allowed the unburnt mixture to be visualised by laser scattering. Consumption of the olive oil by the propagating flame allowed the differentiation of the burnt/unburnt interface.

A schematic of the TCDPIV setup is shown in figure 1 The system used for the characterisation of the flow field/flame interaction consists of two independent DPIV systems. The system was set so both DPIV units imaged exactly the same 80 mm by 80 mm region within the combustion chamber allowing a sequence of two velocity fields to be generated for the instantaneous event. Each DPIV unit comprised of a twin oscillator Nd:YAG laser which produced 50 mJ pulses of 6 ns duration formed into a thin laser sheet 90 mm by 1 mm and a twin framed CCD camera with 1000 pixel by 1016 pixel resolution.

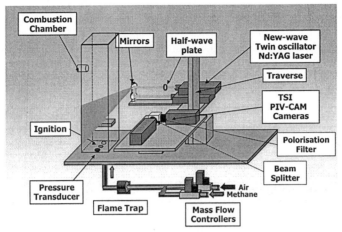

Figure 1 A schematic representation of the explosion chamber and the twin camera DPIV system

Olive oil particles (1-2 μm) where introduced into the combustion air to act as flow tracers and the system was triggered from the mixture ignition. For the data presented, a time separation of 25 μs between frames in each DPIV set and 200 μs between sets was employed.

Analysis of the recorded DPIV images was provided by cross correlation of each image pair over a 32 by 32 pixel interrogation region using a two dimensional FFT algorithm. A 50% overlap was incorporated into the analysis and provided 2.7 mm square interrogation regions spaced at 1.35 mm. The images obtained allow analysis over a 16 by 16 pixel interrogation region but for clarity the above analysis has been presented.

3 Results and Discussion

The overall flame propagation in the combustion chamber is described in figure 2, the images represent a sequence extracted from the HSLSFV recording.

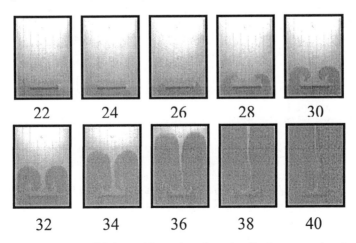

Figure 2 Sequence of high speed laser sheet flow visualisation mages showing flame propagation over the obstacle

The main features of the flame propagation are as follows. As the flame is initiated it propagates out from the central ignition point forming a flame laminar in nature. The flame continues to propagate towards the obstacle pushing the unburnt mixture ahead of it. The movement of the flame front causes a flow field to be generated around the obstacle as the unburnt mixture is forced through the constrictions imposed by the obstacle and containing wall. Turbulent vortex shedding from the obstacle edge occurs as the flow is accelerated. The eddy structures formed are convected downstream into the stagnant wake region behind the obstacle. The flame is also accelerated through the constriction, and on exiting is decelerated as it burns into the wake region. Interaction with the turbulent structures occurs as the flame wraps into the wake region increasing in surface area. Eventually the flame effectively reconnects leaving a small trapped volume of unburnt gas on the downstream obstacle surface.

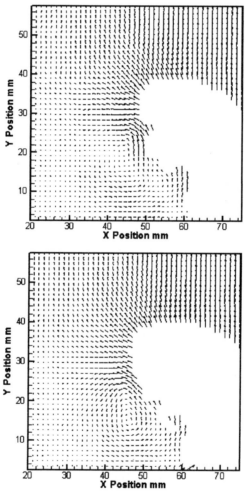

Figure 3 PIV results showing the velocity field in the obstacle wake region at two instances. Separation between images is 200 µs

Figure 3 represents the velocity field for the unburnt mixture in the wake of the obstacle at two instances (separation 200 µs) as the flame begins to wrap into the turbulent structures generated. The cross correlation DPIV results show an approximately constant velocity of 7.5 ms^{-1} for the reactants ahead of the flame and a clear recirculation located in the lower central region of interest. The overall recirculation generated appears to follow the bulk motion of the small scale turbulent eddy structures shed from the obstacle edge. Recirculating at approximately 3 ms^{-1} the flow structures appear to dictate the flame motion and shape in the wake region. In general the flame is burning with the direction of the overall flow, but smaller structures not resolved by the DPIV analysis appear to be interacting with the flame surface.

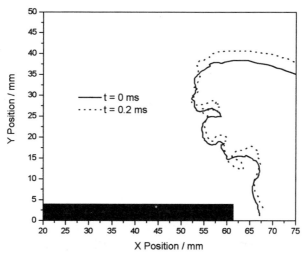

Figure 4 Flame front locations at a time separation of 200 µs in the wake region of the obstacle. Initial flame front approximately 28 ms after ignition

A threshold and edge detection routine was used to define the flame front location at both instances. An example of flame front location for thee results presented in figure 3 is shown in figure 4, highlighting the difference in flame shape over the 200 µs interval. The radius of curvature of the flame front appears different in the central region of each flame, this is due to the recirculating flow structures the flame encounters.

Incorporating the measured velocity field and flame front location, values of flame displacement speed around the flame surface could be calculated. Assuming a local flame displacement normal to the initial flame front a velocity between the two flame front locations at 56 points could be determined. Matching DPIV velocity measurements with points on the initial flame front and resolving the flow velocity in the direction of the local flame velocity values for $V_r.\bar{n}$ were generated. Flame displacement speeds around the flame front, based on the equation described by Sinibaldi et al (1998), are presented in figure 5 ratioed to unstretched laminar flame speed. Values of S_d are shown up to 10 times that of S_{LO} which is larger than those measured by Sinibaldi et al (1998) in a thermally diffusive unstable flame of equivilence ratio 0.6. Negative values of S_d seven times unstretched laminar flame speed are also presented showing good agreement with results produced by Gran et al (1996) in regions of positive curvature.

8

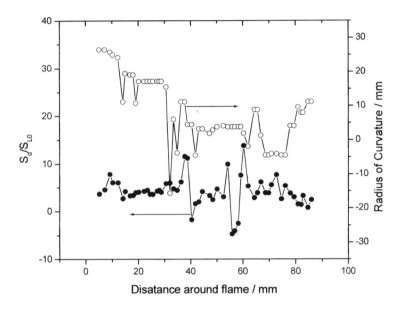

Figure 5 Variation with curvilinear distance around flame front of normalised
flame displacement speed and radius of curvature

4 Conclusions

Twin camera digital particle image velocimetry together with high speed laser sheet flow
visualisation has been shown to provide experimental data for the quantification of flame/flow
interactions in flame propagation studies. A new technique for calculation of flame
displacement speed for a premixed flame interacting with a single rectangular obstacle has
been developed. Measured flame displacements speeds showed values as high as 10 times that
of the unstretched laminar flame speed and negative values of flame displacement speed in
agreement with Gran et al (1996).

Data was shown to be accurately determined in the early stages of flame/vortex
interaction, but unsteadiness in the flow field and uncertainties in the actual flame path can
cause errors in the derived flame displacement speed.

References

Fairweather, M., Ibrahim, S.S., Jaggers, H.and Walker, D.G., 1996, 26th Symp. (Int.) on
Combustion,   365.
Hargrave, G.K. and Williams, T.C., 2000, 9th Int. Symp. of Flow Visualisation, Edinburgh,UK.
Hargrave, G.K., Ibrahim, S.S., Williams, T.C. and Jarvis, S., 2001, Journal of Visualisation.
Hargrave, G.K., Jarvis S. and Williams T.C., 2002, Meas. Sci. Technol. 13, 1.
Lindstedt, R.P. and Saktitharan, V., 1998, Comb. Flame, 114, 469.
Moen, I.O., Donato, M., Knystautas, R., Lee, J.H., 1980, Comb. Flame, 39, 21
Sinibaldi J.O., Mueller C.J. and Driscoll J.F., 1998, 27th Symp. (Int.) on Combustion, 827
Gran I.R., Echekki T. and Chen J. H., 1996, 26th Symp. (Int.) on Combustion, 323

Inst. Phys. Conf. Ser. No. 177
Paper presented at 1st Int. Conf. on Optical & Laser Diagnostics, London, 16–20 Dec. 2002
©2003 IOP Publishing Ltd

Fibre optic sensor arrangements for the analysis of flame propagation in standard spark ignited multicylinder engines

Ernst Winklhofer
Christian Beidl,
Harald Philipp,
Walter Piock

AVL List GmbH
Hans-List-Platz 1
A – 8020 Graz

Abstract. A passive optical method has been developed to enable the inspection of flame propagation events in combustion chambers of spark ignited engines.

The optical system accesses the combustion chamber via individual optical fibers. Each fiber transmitts the flame radiation present within its aperture, the optical signal is converted with photodiodes and is continuously recorded. For the purpose of monitoring local flame radiation properties such as flame front propagation, or localised combustion events such as diffusion burn in pool fire flames or self ignition due to knock or glow ignition, multichannel sensor arrangements are used. The multichannel arrangements are designed to have very narrow aperture cones for each individual optical fiber and geometric arrangements of the fibers which enable the reconstruction of local combustion events. These geometric arrangements cover applications for tomographic reconstruction of flame distribution and for model based event reconstruction if low resolution arrangements must be used. The fiber sensor arrays are implanted in engine components such as the cylinder head gasket or the body of spark plugs. This ensures that the sensors are not compromising the operation of a standard, multicylinder engine.

The paper gives an overview of the field of application and presents application examples for engine engineering.

The topics

Engine development gets its momentum from the availability of system components and actuators which take ever more influence on combustion itself. And equally important, but in contrast to ever more technical content and complexities, there is the demand to exploit the features of existing components for the best tuning of combustion processes.

Developing combustion systems and integrating the variabilities for performance and emissions targets, relies upon well established routines for engine indicating and emissions and performance testing. Sensor application as well as data recording and analysis have become routine methods and highly efficient tools in combustion engine development. Test automation procedures have even succeeded in extending routine analysis methods into automated optimisation of component and system operation modes [1].

However, such comprehensive analysis of components and their vast variabilities also highlight the limitations of conventional optimisation strategies. It is first of all the increasing degree of freedom in finding viable solutions. We strongly believe that finding reliable solutions is only achieved if the variables' effect on combustion events is understood in detail. And it is furthermore that increasingly challenging development targets may overstretch the accessable potential of combustion systems. Both tasks in combustion system development require the thorough understanding of the relevant system mechanisms.

In practical engine development this rises questions about the mechanisms prohibiting the thermodynamic potential of an engine concept to be fully exploited.

- What are the phases in the engine development process where system features are decided?
- And throughout development, when and how is it necessary and appropriate to follow up on this?
- How can all the knowledge gathered in the phase of concept definition and virtual optimisation be effectively rendered in practical development?
- Is our understanding of combustion events in agreement with what is really happening in the cylinder?
- What are the diagnostic tools which enable us to check borderline combustion events in standard, multicylinder engines?

  - How can analysis results be transferred into useful improvements?
  - Are the improvements worth the diagnostics and development effort?

1. Combustion diagnostic techniques

Engine performance, fuel consumption and emissions data are always the start of evaluating a combustion system. This is supported and extended with engine indicating techniques to understand internal mechanisms and thermodynamic status. The outcome of this analysis and the comparison with development targets and with the results of virtual optimisation then defines individual development activities.

This results in specific activities for optimising maximum power density, minimum fuel consumption, especially at part load, and engine torque especially at low and high engine speed. These are extreme conditions for the combustion system, and optimising combustion for these targets finally requires the tuning of flame properties and taking influence on how the flame kernel can spread and propagate through the mixture for best combustion timing and complete consumption of the mixture. How to do this is in general derived from development experience and is more and more supported by numerical optimisation. It all is based on the properties of turbulent premixed flames under the influence of in-cylinder gas motion and gas composition. And in theory, development engineers are aware of how to set engine parameters to take influence on this. But their work could gain considerable precision and efficiency if there is direct feedback about how the flame is in detail responding to variations of combustion system components or operating conditions.

This task of flame visualisation, initially available in research engines only, is now made possible on standard multicylinder engines with sensors and recording techniques appropriate to deliver comprehensive insight into flame propagation and combustion events. A definition of tasks, starting with direct flame imaging in research engines, and a review of techniques and experiences gained from applying AVL's Visiolution techniques in standard engines is given in the following.

2. Flame imaging

In concept studies and in engine pre-development there is a long tradition to use research engines with large windows for the optical access to the combustion chamber. This allows application of very general optical imaging techniques and it yields high resolution flame images as shown in Fig. 1. Flame shape, its position and distribution across the combustion chamber provide the information relevant to understand flame propagation under actual engine conditions on a cycle by cycle basis. As is seen in Fig. 1, this flame shape is best attainable by the direct view through unburned mixture onto the turbulent flame front surface. As soon as the flame front attaches to the window, the view through unstructured radiation of the burned gas dominates the image, which in itself is of very limited information. Essential information, however, may then still be gained from post-flame-front combustion anomalies such as diffusion burn in cases of insufficient mixture preparation.

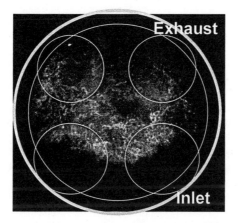

Fig. 1: Observing a premixed turbulent gasoline flame with a CCD camera through a window in the piston of a research engine.

Standard engines can be equipped with small size windows to allow optical access for endoscopic observation. These windows may either be drilled into the cylinder head, or they are part of specially designed spark plugs. Endoscopes together with powerful cameras provide a large field of applications for the direct observation of the events inside the combustion chamber.

An example of spray observation in a DI gasoline engine is given in Fig. 2. With an image intensified camera, this view through the spark plug endoscope allows also observation of the flame kernel. As soon as the flame starts to cover the endoscope window the view onto the front flame surface is lost and the radiation of the burned gas volume reduces image contrast. Then following the further growth of the flame as it expands into the combustion chamber and identifying specific and relevant flame propagation parameters becomes ever more difficult.

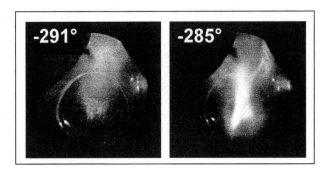

Fig. 2: DI gasoline engine. Spray illuminated and observed through
an endoscope window in the body of a spark plug. Injection at intake stroke.

3. Flame tomography

The method of optical flame radiation tomography was introduced in 1995 [2]. As the methodology has proven to be reliable both in its practical application on the engine as well as in providing relevant information for engine engineering, it has become part of the diagnostic toolbox in developing challenging and advanced combustion systems. Applications are described in references such as [3] and [4]. Concentrating the further refinement of the instrument on areas relevant for combustion system engineering and on optimising sensor units, signal recording and data reduction procedures for best practical use, has resulted in the design of a family of measurement instruments for the combustion diagnostics market [5].

The sensor elements are made of groups of discrete optical fibers and front lenses which form an optical observation grid all across the gasket plane of a gasoline engine combustion chamber (Fig. 3A). The flame radiation intensity within the aperture cone of each fiber is recorded with broadband photo diodes. Simultaneous and continuous recording of up to 160 optical channels and the geometric arrangement of the optical grid form the basis for the numerical reconstruction of flame intensity distribution in the plane of observation.

The actual diagnostic task then requires this data base to be further analysed for flame front propagation, for anomalous flame intensity distribution or for the initiation and propagation of hot spot pre-ignition or knocking combustion.

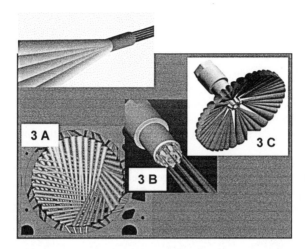

Fig. 3: The core element for Visiolution sensors and standardised sensor configurations.

The spatial resolution of the flame tomography technique is based on the number and distribution of the individual optical channels. It is limited to a few (3 – 5) millimeter and thus by orders of magnitude below the sub-millimeter resolution of direct imaging systems.

However, as the sensors are distributed all around the combustion chamber periphery, there is best use of this resolution throughout all phases of combustion. And in particular, it was found that the important phase of the flame approaching the combustion chamber periphery is recorded with the best possible quality. This has proven to be an indispensable requirement for the precise and reliable analysis of combustion events.

4. Spark plug sensor for flame observation

The fiber optic sensor elements of Fig. 3 can also be arranged in the body of spark plugs, see Fig. 3B. Such fiber optic spark plugs have been in use for many years [6, 7] for the purpose of observing flame radiation as the flame kernel expands beyond the aperture of the optical sensor elements. The signals are used to monitor the time of flame arrival in each optical channel and therefrom to derive the radial growth of the flame kernel. This has proven useful in understanding flame kernel stability or fluctuations and preferential kernel propagation as operating conditions, such as spark timing or port and manifold conditions are modified.

As flame front detection in this multichannel arrangement is based on threshold level settings, the analysis procedure requires channel calibration and the sequencing of threshold levels. Such procedures are integral part of the measurement and data reduction procedure in order to ensure unbiased analysis results.

5. Spark plug sensor for knock center location

The configuration of the small front optic elements can considerably widen the scope of sensor applicability. Figure 3C shows an arrangement which allows flame radiation recording in the entire compression volume. As flame radiation is recorded all along the length of the aperture cones, and as these cones in contrast to the tomography arrangements are not overlapping, there is no radial resolution of this sensor arrangement. There is, however, the resolution all around the cylinder circumference. This allows local, predominant combustion events to be localised with respect to their angular appearance.

This angular resolution is used to monitor the direction at which self ignition under knocking conditions occurs. Self ignition of endgas in combustion chambers with a central spark plug is usually confined to areas close to the cylinder liner. Hence the idenfication of the angular position of the self ignition event is sufficient information for determining knock center location.

This task is achieved with a spark plug shown in Fig. 4 (see also refs. [8 and 9]). A single channel signal together with the pressure signal is also shown in Fig. 4. Flame radiation yields an increasing intensity signal from about −10° onwards. It rises with the ongoing combustion and just before the pressure signal identifies knocking combustion, the flame radiation signal already responds to this fast combustion event.

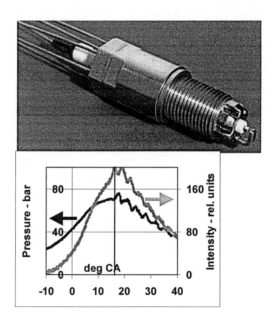

Fig. 4: Spark plug with 40 optical fibers for observation of the compression volume. Pressure signal and flame radiation intensity signal recorded within the acceptance cone of one fiber optic channel.

The rise of the radiation signal is first of course noticed by the sensor channel which is in direct view of this event. Consequently, it is the task of the multichannel data reduction procedure to find the channel which first responds to this high frequency combustion event. In order to corroborate the analysis result, the multichannel high pass filterd signal patterns are analysed with respect to the signal model of Fig. 5.

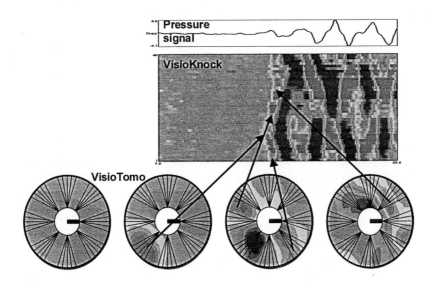

Fig. 5: High pass filtered signals at knocking combustion: Pressure oscillation amplitudes, flame radiation intensity modulation in all 40 optical channels of the VisioKnock sensor and propagation of the „brightness wave" reconstructed from flame tomography measurements. Line-of-sight graphic of the VisioKnock sensor superimposed on the VisioTomo reconstructions.

The results example of Fig. 6 shows the data relevant for identifying the knock center location. The signal pattern of the VisioKnock sensor shows that self ignition is first noticed along the direction of channel nr. 36.

Self ignition at the periphery of the combustion chamber creates a pressure wave and the rising gas density in the front of this wave gives rise to the "brightness" wave of the brightly radiating burned gas. As this brightness wave propagates across the combustion chamber it covers more and more of the neighbouring channels.

Knock pressure amplitudes may be as small as 1 or 2 bar and still yield high frequency signal patterns which allow the simple and reliable identification even of boarderline knock events.

16

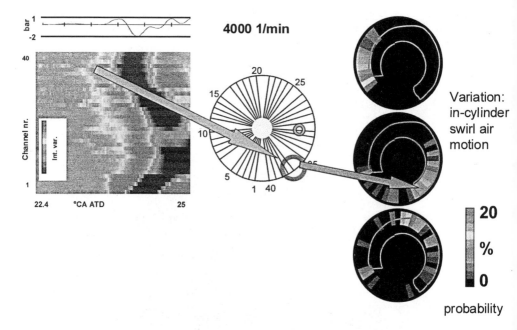

Fig. 6: VisioKnock: Single cycle result and knock center distributions
derived from multicycle datasets.

Using such spark plug sensors in test procedures, either on the engine test bed or at the
chassis dyno, requires sensor reliability and the matching of original spark plug geometry
and heat value. These quality requirements are achieved by adequate selection of sensor
material and manufacture processes.

6.   Integrating Visiolution techniques into engine development tasks

In engine development projects, the need for information about flame properties arises as
soon as engine boarderline conditions do not meet their thermodynamic targets. Particular
topics are idle stability, part load de-throttleing, or highest power density at wide open
throttle operation. First choice for system improvements is optimisation of operating
parameters. If this appears to be insufficient, there is the requirement for modifications of
components and system design and the necessity to integrate such modifications into
combustion system development. As the effort for such hardware iteration loops can be quite
considerable, the careful preparation of test variants is of utmost importance.

6.1   Instantaneous combustion diagnostics

Quite often, an on the spot demand for flame diagnostics arises out of actual development
priorities. A quick reaction to such demands is provided with off the shelf spark plug
sensors.

Measurements are done either on the engine test bed or in the vehicle on a chassis dyno as shown in Fig. 7. And given the sensor and accesories design, there is not even the need for in advance engine preparations.

Fig. 7: VisioKnock application in a vehicle on the chassis dyno.

The application of the VisioKnock sensor is especially focused on the identification of knock centers along the engine's full load line. In addition to this primary focus, the practical work with the sensor has shown that it is also suitable for identifying combustion anomalies which appear on top of the turbulent flame propagation in ideally homogeneous premixed charge.

Thus, the inspection of flame radiation intensity traces can pinpoint anomalously bright combustion events such as pool fire diffusion flames. An example is given in the lower part of Fig. 8 showing the time traces of flame intensities along the 40 radial observation cones around the combustion chamber. The significantly enhanced flame brightness found in the sectors around the intake valves is an indicator for sooting diffusion burn in this area. This is obviously the flame signature following imperfect mixture preparation near the intake valves.

The early phase of combustion is inspected with the VisioFlame sensor. Information is provided on preferential flame kernel propagation, either on a cycle by cycle or on a statistical basis.

A summary of results gained with the spark plug sensors for flame kernel inspection, knock center distribution and diffusion flame areas is given in the polar diagrams of Fig. 8.

18

The spark plug sensors are suited for application in the entire speed and load range of modern engines.

Especially the VisioKnock sensor is continuously used along the full load line under knocking conditions. The simple mounting procedures allow for consecutive cylinder screening, or simultaneous measurement in different cylinders. This opens a wide field of applications for cylinder variation analysis and even for questions on combustion events which may be correlated between groups of cylinders.

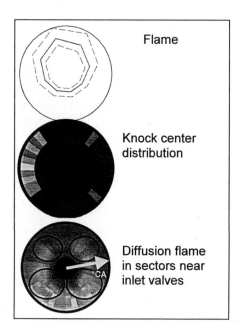

Fig. 8: Examples of measurement results from the spark plug sensors:
Flame kernel pattern, knock center distribution, bright diffusion
flames in the intake valve sectors between 15 and 40 °CA ATDC.

6.2 Systematic preparation of combustion analysis

Developing modern combustion systems relies upon virtual optimisation in the design phase and on hardware testing with best standardised indicating and visualisation procedures. Flame tomography provides here the most comprehensive means for flame analysis under real engine conditions.

Hardware preparations for flame tomography comprise of manufacturing the engine specific gasket sensor and of maching the block and head of the engine to provide the space necessary for the gasket plate with the sensor array.

Tomography results are derived from data reduction procedures. They are usually presented with plots of flame front propagation patterns, and in the case of knock analysis, with knock center distribution statistics as shown in Fig. 9 for a four valve combustion chamber.

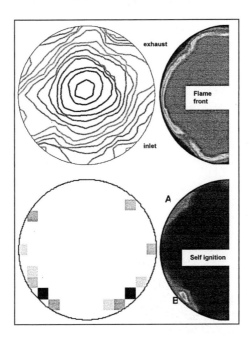

Fig. 9: Flame tomography in the multicylinder engine with data reduction
for flame front propagation and knock center distribution.
Combustion modeling: Full load, 15°CA ATDC.

The data show that from the initial combustion phase onwards the flame has a slight preference towards the exhaust valves. It then approaches the cylinder liner in a non uniform way, with considerable flame stretch towards the neighbour cylinders and retarded propagation into the intake side.

This particular operating point has also been studied in a CFD analysis. A flame front map and end gas reactivity distribution are shown in Fig. 9 for the time instant at 15 °CA ATDC. As a consequence of symmetrical boundary conditions, the modeling results are symmetric and they are displayed just for one half of the combustion chamber.

The comparison with tomography data shows the agreement with the flame stretch effects, and end gas reactivity modeling precisely identifies the areas which were later experimentally found to be centers for self ignition.

Where then is the benefit of VisioTomo measurements when the results are as well predicted by modeling ?

Reliable modeling results for boarderline operating conditions are only achieved, if in addition to appropriate models, the relevant boundary conditions are understood and correctly implemented. For engine development, this raises the essential question on how sensitive a result depends on variations of these geometric or operating conditions and which of these conditions are relevant for engine behaviour.

A geometric condition can be the depth position of the spark electrode. For the same engine as above, the data in Fig 10 show that, with a deeper spark position, the flame first propagates towards the intake side. The endgas at the exhaust side then has more time to build up reactivity and, consequently, knock centers are now accumulated near the exhaust.

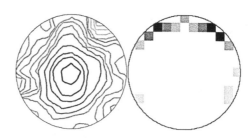

Fig. 10: Long spark plug with deep electrode position: Advanced flame propagation towards inlet side, knock centers accumulate on the exhaust side.

The mechanism for this particular behaviour is found as we check the response of the flame kernel towards the influence of the convective airmotion. In the tumble flow field shown in Fig. 11, the flame kernel introduced near the cylinder head with a flat spark plug is pushed towards the exhaust side, whilst introducing the flame kernel with a deep electrode position has it convected towards the intake valves. The overall effect on the flame front propagation patterns may at first not at all be dramatic as the tomography data show, but endgas burnoff is either on one side or on the other. And offering endgas to the high temperature environment at the exhaust side has certainly the higher risk of knocking combustion and of less efficient combustion phasing and IMEP levels than doing so for the intake side.

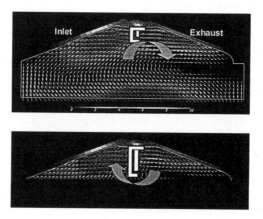

Fig.11: Explaining the results in figs. 9 and 10: depending on electrode position, the tumble flow convects the flame kernel to either the outlet or intake side.

Even if best possible CFD results are available, examples like the above demonstrate that the effort necessary to yield flame diagnostic data in fully operative engines is far from redundant. Quite in contrast, it becomes evident that the information prepared by modeling studies receives substantial focus as soon as flame data from the actual engine are available.

This is especially found for all "boarderline" conditions such as knock or stability limits at lean and high EGR operation. And quite often, the importance of otherwise unrecognised geometric or operative details and their influence on the success of the system only becomes evident from the direct feedback out of the combustion chamber. In combustion diagnostics projects it was also found that the effort for additional variants studies following the baseline analysis is rather small, the benefits, however, to be gained for the interpretation of the systems' behaviour and consequently for reduction of test bed time, can be substantial.

7. Summary

This article presents diagnostic techniques for flame analysis in multicylinder engines. The sensors allow unrestricted operation of the engine in its entire speed and load regime. Access to the combustion chamber is achieved with optical fibers together with micro-optic front elements which are implanted in spark plug bodies or in special metal layers between cylinder head and block.

The fiber optic sensors have been designed for the analysis of flame propagation events which are relevant for combustion system development in standard multicylinder engines. As the sensors provide direct insight into flame propagation events, the results should support the understanding of local combustion events under the influence of combustion system components and component operation modes.

The experience gained from sensor applications in numerous analysis projects both on the engine test bed as well as on the chassis dyno, has confirmed the robustness of sensor technology and the maturity of data recording and data reduction techniques. The request to record combustion events with a degree of spatial and temporal resolution, appropriate to identify the potential for combustion system improvements is met with a series of standardised sensors and data reduction procedures.

Combining the optical analysis with results derived from conventional engine tests and indicating as well as the analysis support available from numerical analysis has proven to provide the diagnostic precision required for finding solutions in challenging development projects. This precision and the immediate analysis feedback to variations of component properties reduces iteration loops in engine development and allows the focusing of development efforts onto most promising options.

References

[1]   K. Gschweitl, H. Pfluegl, T. Fortuna, R. Leithgoeb: *"Steigerung der Effizienz in der modellbasierten Motorenapplikation durch die neue CAMEO Online DoE-Toolbox"*, ATZ Automobiltechnische Zeitschrift 103 (2001) 6 p2-7

[2]   H. Philipp, A. Plimon, G. Fernitz, A. Hirsch, G. Fraidl, E. Winklhofer: *"A Tomographic Camera System for Combustion Diagnostics in SI Engines"*, SAE 950681

[3]   E. Winklhofer and G. Fraidl: *"Optische Indizierverfahren für Benzin DI Verbrennungssysteme – Aufwand und Nutzen"* ("Optical methods to assist the development of GDI combustion systems – efforts and benefits") , MTZ Motortechnische Zeitschrift, November 1998

[4]   J. Liebl, J. Poggel, M. Klüting, S. Missy: *„Der neue BMW Vierzylinder-Ottomotor mit Valvetronic"*, MTZ Motortechnische Zeitschrift 62 (2001) 7/8 MTZ 2001

[5]   www.avl.com/visiolution

[6]   U. Spicher, G. Schmitz, H.P. Kollmeier: *„Application of a new optical fiber technique for flame propagation diagnostics in IC engines"*, SAE 881647

[7]   P.O. Witze, M.J. Hall, J.S. Wallace: *„Fiber-optic instrumented spark plug for measuring early flame development in spark ignition engines"*, Transactions of the SAE 97, p.3.813

[8]   H. Philipp, A. Hirsch, M. Baumgartner, G. Fernitz, Ch. Beidl, W. Piock, E. Winklhofer: *"Localisation of knock events in direct injection gasoline engines"*, SAE 2001-01-1199

[9]   E. Winklhofer, Ch. Beidl, H. Philipp, W. Piock: *"Optische Verbrennungsdiagnostik mit einfach applizierbarer Sensorik"*, MTZ 9/2001 p 644 – 651

Inst. Phys. Conf. Ser. No. 177
Paper presented at 1st Int. Conf. on Optical & Laser Diagnostics, London, 16–20 Dec. 2002
©2003 IOP Publishing Ltd

Planar Laser Induced Fluorescence Measurements of Fuel Concentration in a Gasoline Direct Injection Optical Engine

Sara Gashi, Youyou Yan, Russel D Lockett and Constantine Arcoumanis

Centre for Energy and the Environment, School of Engineering and Mathematical Sciences, City University London

Abstract. The Planar Laser Induced Fluorescence (PLIF) technique was used to characterise the fuel concentration distribution inside a single-cylinder direct-injection gasoline optical engine, in terms of two-dimensional, instantaneous and high-resolution images. The research engine employs a 4-valve pent-roof cylinder head, with the optical access obtained through both a pent-roof and an elongated piston window. The multi-hole injector and the spark plug are centrally located in the cylinder head, generating a spray-guided system which represents the basis of second-generation direct injection engines.

The research engine was fuelled with a mixture of iso-octane and a fuel tracer; two different tracers, 3-pentanone and TEA (triethylamine), excited by an Nd:YAG laser at 266nm were investigated. The fuel was injected late in the compression stroke, aiming to provide a stratified air/fuel mixture near the spark plug at the time of ignition. The degree of charge stratification was characterised by qualitative and quantitative measurements of the crank-angle resolved liquid and vapour fuel concentrations in the form of equivalence ratio distributions.

1. Introduction

The continuing development of gasoline direct-injection engines is under consideration in the automotive industry in Europe and Japan, due to disatisfaction with the first-generation engines and due to increasing competition from direct injection diesels. The advantages of adopting the direct-injection gasoline concept, because of its outstanding potential for fuel economy and CO_2 reduction, have been well documented [1-4] and remain its strongest selling point.

Different configurations of combustion chambers have been designed, corresponding to wall-, air- or spray-guided systems, each with advantages and disadvantages [5]. The configuration adopted in this study is the spray-guided system, using the close spacing approach where the basic aim is to provide consistenly a slightly rich mixture around the spark plug at the time of ignition, by allowing enough time for vaporization of the fuel after injection and preventing wall wetting. In order to realize the maximum possible fuel economy, the engine has to be operated unthrottled. Burning an overall lean mixture in spark ignition engines but with local charge stratification [6], offers a number of advantages but the difficulties remain in achieving a stratified charge over the full regime of low loads and speeds; some degree of throttling may thus be required, resulting in losses under some operating conditions.

Planar Laser Induced Fluorescence (PLIF) is a well established technique for in-cylinder measurements of local fuel concentration and fuel distribution, and it has been widely used for investigations in both port injection and direct injection gasoline engines [7-11].

In this paper we present radial and axial LIF measurements of fuel distribution in a second-generation DI gasoline optical engine. The results obtained give information about the extent of charge stratification, the level of cyclic fluctuations, and the fuel/air mixture formation process, in a way that allows validation of relevant computer codes and improved engine design.

2.   Experimental System

2.1   Optical Engine

The specifications of the optical engine used in this investigation are listed in Table 1. The engine is a single-cylinder one, equipped with a four-valve pent-roof cylinder head, a direct fuel injection (common-rail) system, a flat elongated piston and a displacement of 0.35 litres. The injector and the spark plug are centrally located in the cylinder head, i.e. creating a spray-guided system with close spacing configuration.

Table 1: *Engine specifications*

Engine type	4-stroke DI Gasoline
Combustion Chamber	4-valve, pent-roof
Bore (mm)	75
Stroke (mm)	79.2
Displacement (l)	0.35
Compression Ratio	10.5
Injector type	High Pressure Multi-hole
Injector location	Near Central

The combustion chamber geometry is the same as that of the corresponding production engine. The inlet manifold provides a flow with a significant tumble motion, in order to enhance the turbulence intensity and charge stratification at the time of ignition.

The engine could be run fired or motored. In the test reported here, the engine was motored by a shunt dynamometer. The crankshaft is fitted with two shaft encoders (a low-resolution one providing 1 pulse/revolution and a high-resolution one providing 1800 pulses/revolution) which provide synchronisation signals for the engine control unit (SCECU), that in turn control the injection start timing and duration, and provide control signals to trigger the laser and the camera.

The injector employed was a high-pressure multi-hole injector, with injector pressure of 120 bar. The higher fuel pressure was used to finely atomise the spray in order to vaporize the fuel within the very short timescales required in the late injection mode.

The engine cylinder liner has a 40mm long, 17mm thick fused silica ring fitted immediately under the cylinder head, which enables the laser sheet to pass through the combustion chamber. Optical access through the piston was obtained with a 45-degree mirror mounted just above the engine block inside the elongated hollow piston. The fused silica piston window occupied 61% of the total piston top surface. An additional fused silica window was employed on one side of the cylinder head, in order to provide optical access within the cylinder head to the spark plug and injector tip region. However, as this optical access was available on one side only, any horizontal laser beam entering the cylinder head could not exit with unavoidable scattering/reflection from the opposite metal surface.

The in-cylinder pressure was measured with a Kistler water-cooled piezo-electric pressure transducer, with a sensitivity of 26.5 pC/bar. The pressure trace was recorded using AVL 620 Indiset system.

2.2   LIF System

A Spectra Physics Lab-170 Nd:YAG laser operating at 10 Hz and 12.5Hz repetition rate in external trigger mode, corresponding to engine speeds of 1200 rpm and 1500 rpm respectively, was employed; it produces 266nm pulses of 92 mJ/pulse maximum energy and 10 ns duration. The laser energy used for the experiments was 80mJ/pulse for radial LIF (RLIF) and 40 mJ/pulse for axial LIF (ALIF). An electronic signal from the engine control unit (SCECU), which also controlled the timing and duration of the injection, was employed to trigger the laser and the camera.

The laser beam was formed into a collimated laser sheet with a 6x cylindrical telescope (-25mm and +150mm focal length lenses), and focused with a 1 m focal length spherical lens to form a laser sheet with thickness of 0.1mm at focus at the centre of the combustion chamber. The edges of the laser sheet were truncated, producing a laser sheet of approximately 30 mm width.   For RLIF the laser sheet entered the combustion chamber through the glass liner, and the images were captured with the camera viewing from the 45 degree mirror through the window in the piston. For ALIF the laser sheet entered the combustion chamber from the 45 degree mirror through the piston window, and was centrally aligned with the injector. The images were captured at 90 degrees to the laser sheet, as shown in Fig. 1, using a DiCAM-Pro PCO 12bit Intensified CCD camera, with a spatial resolution of 1024x1240 pixels.

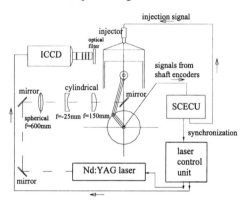

Figure 1 Schematic Diagram of ALIF Set-up

The maximum frame rate of full image recording is 8 Hz. This rate increases as the image size decreases. In this test, the image was binned 2x2, therefore the frame speed was high enough to capture images of consecutive cycles at an engine speed of 1500rpm. For the TEA fluorescence collection the camera was equipped with a UV – 94mm f/4.1 fused silica lens, and a 280nm long -pass filter was used to reject any elastic scattering at 266 nm. For collecting fluorescent light from 3-Pentanone, a Nikon f/2 135mm lens was used together with a 365-435nm band pass filter. The 3-Pentanone fluorescence has a peak intensity at approximately 410 nm when excited at 266 nm. The band pass filter was used to filter out any light from elastic scattering and fluorescence at other wavelengths.

2.3   Fuel Tracers

The engine was fuelled with pure Iso-octane which does not normally absorb any laser light at 266nm wavelength and does not fluoresce, necessitating the addition of a fuel tracer to the fuel. Two different fuel tracers were employed and compared in this study: 3-Pentanone [7,10,11] and Triethylamine (TEA). TEA has been most often used as an exciplex agent [12,13]. When employed as a fluorescent fuel marker in engine LIF experiments, it exhibits a slightly different fluorescence mechanism than 3-Pentanone. When 3-Pentanone is used as a fluorescence tracer, its fluorescence intensity is proportional to the number of 3-Pentanone molecules in the laser sheet, and therefore the fluorescence intensity provides a measure of the molecular or molar concentration of 3-Pentanone, enabling the determination of the fuel molecular or molar concentration. On the other hand, the fluorescence intensity of TEA is proportional to the ratio of the number of TEA molecules to the number of Oxygen molecules. Therefore, the fluorescence intensity provides a direct measure of the fuel-air ratio and/or equivalence ratio.

Fluorescent fuel tracers should satisfy a number of physical and chemical requirements. They should be soluble in the fuel, and have a similar boiling point to ensure representative vaporisation and mixing. They should have a molar mass comparable to that of the fuel, to ensure that the fluorescent tracer follows faithfully the fuel during any diffusion controlled processes. The fluorescence light should exhibit minimum pressure and temperature dependence. Fluorescence obtained from 3-Pentanone in the vapour state has minimal pressure dependence, and a linear dependence on temperature [8]. The physical properties of both tracers are listed in Table 2 along with those of iso-octane.

Table 2: *Properties of fuel and tracers*

	Iso-octane C_8H_{18}	3-Pentanone $C_5H_{10}O$	Tri-Ethyl-Amine (TEA) $(C_2H_5)_3N$
Molecular Weight (g/mol)	114.23	86.1	101.19
Density (g/cm³)	0.688	0.816	0.724
Boiling Point (K)	372.3	374.65	362.6
Enthalpy of Vaporization (kJ/mol)	35.15	38.2	35.17

3. Image Processing

The fluorescence intensity is proportional to either the molecular concentration (3-pentanone) or the equivalence ratio (TEA), depending on which fluorescence tracer is used. The quantitative measurement of the molecular concentration or the equivalence ratio requires a reference point at which the measured quantity should be known. In this experiment, the calibration for quantitative measurement was conducted by generating a homogeneous air-fuel mixture in the cylinder. A port injector was used to inject fuel before the air intake manifold during the induction stroke which allows a homogeneous fuel distribution to be generated in the engine cylinder at the end of the compression stroke. The mass flow rate of fuel and air were measured with a fuel metering system and an air drum, respectively. Hence the molecular concentration or the equivalence ratio of the homogeneous air-fuel mixture field could be accurately determined.

A number of images were taken at each specified crank angle for calculating the mean and the standard deviation. Single shot images were processed by using the following equation,

$$C_{ij} = cS_{ij}^{D} = c \frac{R_{ij} - c_1 \overline{B}_{ij}}{L_{ij} - c_2 \overline{B}_{ij}^{L}} \quad ......(1)$$

where R is the single shot image of the fuel spray from the injector under investigation, $\overline{B}$ is the mean of the background images obtained from the motored engine with no injected fuel, L is the mean of the homogeneous fuel field images created by injection during induction in the intake manifold, and $\overline{B}^{L}$ is its mean background image. Constants c_1 and c_2 were introduced into the equation to reduce the effect of secondary scattering, which proved significant in the RLIF images. In the ALIF, c_1 and c_2 were both set equal to 1, since secondary scattering was negligible in most of the area of the images. Constant c was determined as the ratio of imaging parameters between test and its calibration, i.e. the ratio of camera lens aperture, intensifier gain and reduction coefficient of a neutral density filter used in front of the camera.

The final images of 3-Pentanone fluorescence distribution shown in the next section were processed as described above. For the TEA fluorescence images, the final images were multiplied by the ratio of the stoichiometric AFR over the calibration AFR; therefore the final images represent the 2-dimensional distribution of the equivalence ratio across the measurement plane.

4. Results and discussion

The engine test conditions are listed in Table 3. For all the experiments the throttle remained wide open (WOT). The fuel injection into the cylinder occurred at 120 bar pressure, for a duration of 1.15 ms, which corresponds to an overall cylinder AFR of approximately 60.

The homogeneous fuel concentration field used for calibration was generated using a single-hole port injector. The injection duration was 5ms at 120 bars and the mean AFR of the homogeneous fuel field was 6.3. the multi-hole nozzle of the high pressure injector created a spray pattern simulating diesel sprays; a typical example is given in Fig.2 for injection under atmospheric conditions.

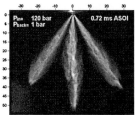

Figure 2 Spray structure generated by the multi-hole injector

Table 3: *Engine operating conditions*

	RLIF	ALIF	ALIF
Engine Speed (rpm)	1500	1500	1500
Injection Pressure (bar)	120	120	120
Injection duration (ms)	1.15	1.15	1.15
Start of Injection (BTDC)	45	40	40
Fuel and Tracer	90%Iso-octane 10% 3-Pentanone	80%Iso-octane 20% 3-Pentanone	95%Iso-octane 5% TEA
Imaging CAD (BTDC)	28, 30, 32, 36, 38	20,22,24,26 28,30,32,34,35	20,22,24,26 28,30,32,34,35

4.1 Radial LIF

The laser sheet used for RLIF was inclined about 2^0 to the horizontal (Figure 3) and entered the engine through the pent roof window, while exiting through the optical liner (in the absence of an optical window at the other side of the pent roof). The laser sheet was placed

28

at about 10 mm under the injector tip, which was centrally mounted at the top of the pent roof.

Electronic start of injection occurred at 45 deg. BTDC. There was a delay of 5 deg. CA between the injection signal and the first appearance of fuel at the tip of the injector. Since the laser sheet was approximately 10 mm below the injector, the fuel spray intersected the laser sheet at 38^0 BTDC, i.e. 2^0 CA later. The latest imaging angle possible using this arrangement was 28^0 BTDC, after which the top of the piston intersected the plane of the laser sheet.

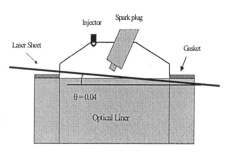

Figure 3 Laser sheet in the RLIF

The ambiguity between the liquid and vapour phases, of the fuel spray can be resolved by the sharp change in fluorescence intensity; that of the liquid phase at 32^0, 36^0 and 38^0 BTDC is much higher than that at 28^0 and 30^0 BTDC. In resolving the spray structure, there are always vaporised fuel clouds around the liquid sprays, but it is too difficult to estimate the fuel vapour concentration simultaneously with the liquid phase.

Images of vapour phase fluorescence intensity distribution processed by equation (1) are proportional to the molecular fuel concentration. Images of liquid phase fluorescence were processed in the same way as the vapour phase images, but they are not a direct measurement of fuel concentration. Nevertheless, the processed images may be used as qualitative visualisation of liquid phase fuel concentration.

Figure 4 shows images of the liquid and vapour phases at 38^0 BTDC and 28^0 BTDC, respectively. For each crank angle, the image set includes 2 single shot images and the mean and standard deviation over 69 images.

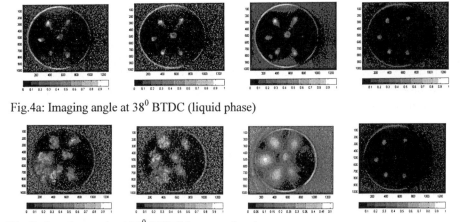

Fig.4a: Imaging angle at 38^0 BTDC (liquid phase)

Fig.4b: Imaging angle at 28^0 BTDC (vapour phase)

All images in Fig.4 correspond to HLIF with 3-pentanone; from left to right (i), (ii) single shot images, (iii) mean image and (iv) standard deviation image.

The single shot and the mean images show the spatial distribution of the 6 jets exiting from the injector. The change in the single shot images represent the cyclic variation between shots and is measured by means of the standard deviation.

4.2 Axial LIF with 3-Pentanone

The relevant position of the ICCD camera to the injector is shown in Fig. 5.

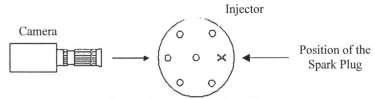

Figure 5 Camera position in ALIF

It is using this viewing angle, that the images in Fig. 6 were obtained. Figure 6 shows in terms of the spray evolution as a function of crank angle the mean images over 50 single shots. The start of injection occurred at 40^0 BTDC, and the spray first appeared at 35^0 BTDC. There is a sharp change of fluorescence intensity occurring between 26^0 and 28^0 BTDC; images after 26^0 BTDC are proportional to the molecular fuel concentration, whereas images before 28^0 BTDC are simply a qualitative visualisation of the liquid fuel spray.

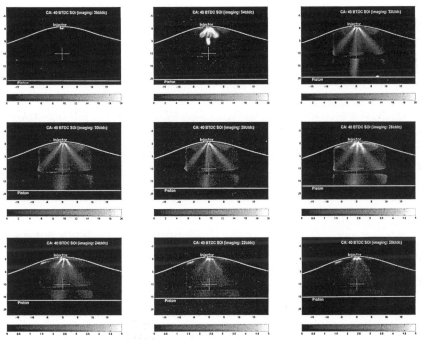

Figure 6 ALIF (fuel tracer 3-pentanone), mean over 50 images, imaging crank angles: 35, 34, 32, 30, 28, 26, 24, 22 and 20 deg BTDC

The spray penetration distance, the single spray angle and the angle between individual sprays can be estimated from these images. Spray impingement on the piston surface is observed after 32^0 BTDC.

4.3 Axial LIF with TEA

Figure 7 presents an example of the equivalence ratio distribution at one crank angle obtained by means of TEA ALIF. The camera position is the same as that in the ALIF with 3-Pentanone (Figure 5). The start of injection was 40^0 BTDC, and images were taken at 9 crank angles (the same as with 3-Pentanone). The sharp fluorescence intensity change occurred at the same crank angles, between 26^0 and 28^0 BTDC.

Figure 7 shows that there is no valid data in the region close to the injector. When TEA is used with the UV-camera lens and the 280 nm long pass filter, light scattering from the spark plug, which was immediately behind the injector, was sufficiently intense to damage the intensifier. An optical mask was employed to eliminate any scattering from the spark plug.

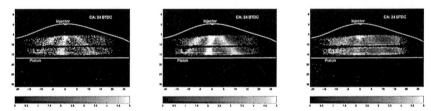

Figure 7 ALIF (fuel tracer TEA), from left to right: a single shot image, mean over 50images, standard deviation over 50images, Imaging angle 24^0 BTDC

The advantage of using TEA as a fluorescent marker is that a processed image by means of equation (1) provides the equivalence ratio distribution which is non-dimensionalised, and more useful for comparison with other experiments and CFD simulations. The molecular fuel concentration can be converted to equivalence ratio, when extra data on pressure and temperature inside the engine cylinder are available [7]. Even so the deduction of the equivalence ratio from fuel concentration will increase the uncertainty in the equivalence ratio measurement.

The maximum equivalence ratio according to Fig. 7 is just over 5, which is significantly higher than the mean equivalence ratio of 0.25 calculated from the mean engine AFR. Thereofore, the engine can run very lean but the local AFR in the imaged region is very rich.

5. Concluding Remarks

The fuel spray produced from a multi-hole injector was investigated in a single-cylinder, direct-injection gasoline optical engine by Planar Laser Induced Fluorescence. Measurements were obtained along a radial and an axial plane relative to the geometry of the engine cylinder and used to characterise the spray produced by the multi-hole injector, in terms of spray cycle-to-cycle stability, spray penetration and fuel distribution.

Two fluorescence tracers were tested: 3-pentanone and TEA. Processed fluorescence images with 3-Pentanone determined the distribution of the molecular fuel concentration, while the advantage of using TEA is the direct estimation of the equivalence ratio distribution in the engine cylinder.

6. Acknowledgments

The authors would like to thank Dr. K.S. Jeong from the Korean University of Technology and Education, Tom Flemming and Nicholas Mitroglou from City University London, for their assistance at various stages of the experimental programme.

References

[1] Zhao, F.Q., and Lai, M.C. " A Review of Mixture Preparation and Combustion Cnontrol Strategies for Spark-Ignited Direct-Injection Gasoline Engines" SAE paper 970627,

[2] Wirth, M., Piock, W.F., and Fraidl, G.K. "Actual Trends and Future Strategies for Gasoline Direct Injection" S433/005/96 ImechE, 1996

[3] Fraidl, G.K., Piock, W.F., and Wirth, M. "The Potential of the Direct Injection Gasoline Engine" 18th Int. Vienna Motor Symposium, 1997

[4] Eichlseder, H. ,Baumann, E., Muller, P., and Rubbert, S. "Gasoline Direct Injection- A promising Engine Concept for Future Demands" SAE paper 2000-01-0248, 2000.

[5] Ortmann, R., Arndt, S., Raimann, J., Grzeszik,R. and Wurfel, G. "Methods and Analysis of Fuel Injection, Mixture Preparation and Charge Stratification in Different Direct Injected SI Engines" SAE paper 2001-01-0970, 2001.

[6] Arcoumanis, C., Hull, D.R., and Whitelaw, J.H. "An Approach to Charge Stratification in Lean Burn, Spark- Ignition Engines" SAE paper 941878, 1994.

[7] Berckmuller, M., Tait, N.P., Lockett, R.D., and Greenhalgh, D.A. "In Cylinder Crank-Angle Resolved Imaging of Fuel Concentration in a Firing Spark-Ignition Engine using Planar Laser Induced Fluorescence" 25th Int. Symposium on Combustion, The Combustion Institute, pp. 151-156, 1994.

[8] Fujikawa, T., Hattori, Y., and Akihama, K. "Quantitative 2-D Fuel Distribution Measurements in an SI Engine using Laser Induced Fluorescence with Suitable Combination of Fluorescence Tracer and Excitation Wavelength" SAE paper 972944, 1997.

[9] Gold, M., Stokes, J., and Morgan, R. "Air-Fuel Mixing in a Homogenous Charge DI Gasoline Engine" SAE paper 2001-01-0968, 2001.

[10] Vannobel, F., Arnold, A., and Buschmann, A. "Simultaneous Imaging of Fuel and Hydroxyl Radicals in an In-line Four Cylinder SI Engine" SAE paper 932696, 1993

[11] Gold, M.R., Arcoumanis, C., Whitelaw, J.H. "Mixture preparation strategies in an optical four valve port-injected gasoline engine" Int. Journal Engine Research, Vol.1, no.1, 2000

[12] Jeong, K.S., Jermy, M.C., and Greenhalgh, D.A. "Laser Sheet Dropsizing in evaporating sprays using Laser Induced Exciplex Fluorescence" 9th Int. Symposium on Flow Visualisation, Heriot-Watt University, Edinburgh, 2000.

[13] Ipp, W., Wagner, V., Kramer, H., Wensing, M., and Leipertz, A. " Spray Formation of High Pressure Swirl Gasoline Injectors Investigated by 2D Mie and LIEF Techniques" SAE paper 1999-01-049,1999.

Inst. Phys. Conf. Ser. No. 177
Paper presented at 1st Int. Conf. on Optical & Laser Diagnostics, London, 16–20 Dec. 2002
©2003 IOP Publishing Ltd

Acousto-optic frequency switching for two-wavelength planar Doppler velocimetry

Helen D Ford, David S Nobes and Ralph P Tatam

Optical Sensors Group, Centre for Photonics and Optical Engineering, School of Engineering, Cranfield University, Bedfordshire, MK43 0AL, UK.

Abstract. In planar Doppler velocimetry a laser sheet illuminates a seeded flow. A portion of the sheet is imaged, through a cell containing gaseous iodine, onto a solid-state camera. The laser frequency is tuned to coincide with an absorption line of iodine. The frequency of light scattered from the measurement volume experiences a velocity-dependent Doppler shift and the iodine cell acts as a frequency-to-intensity converter, mapping the Doppler shifts as a spatial variation of intensity across the image.

Referencing is required to normalise the intensity-modulated Doppler image. Generally, half the image beam power is split off before the cell to form a reference image on a second camera. However, sufficiently precise pixel-to-pixel matching of the signal and reference images is a major processing challenge, especially when high velocity gradients exist.

The difficulty can be circumvented, for steady-state flows, by illuminating sequentially with two optical frequencies, one tuned to an absorption line and one to a region of no absorption. Thus signal and reference images can be acquired using a single camera, achieving automatic pixel matching. One or more acousto-optic frequency-shifters in our system allow both illumination beams to be derived from a single argon-ion laser. Successful results have been obtained from a rotating disc.

1. Introduction

Planar Doppler velocimetry (PDV) is a flow measurement technique that provides velocity information over a plane defined by a light sheet formed from an expanded laser beam [1],[2]. The optical frequency of light scattered from each particle in the seeded flow experiences a Doppler shift, which is linearly related to the velocity of the particle at that point in the flow. A portion of the light sheet is imaged, through a glass cell containing iodine vapour, onto the active area of a solid state camera. The laser frequency is chosen to coincide with an absorption line of iodine [3] so that the optical intensity at any position in the camera image is a function of the Doppler shift experienced at the corresponding flow position.

The intensity over the image is also modulated by the Gaussian intensity profile of the beam and non-uniform seeding density. Therefore it is usual to split off part of the image beam before the iodine cell [4], to form a reference image on a second camera.

Superposition of the two images must be achieved to sub-pixel accuracy, particularly if large velocity gradients are present in the region imaged. Software de-warping routines can achieve good pixel matching [5] and can help to correct for small differences in

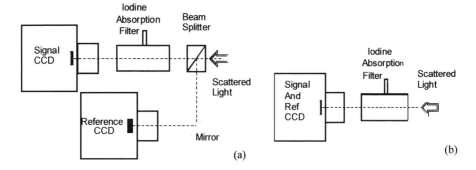

Figure 1. (a) Conventional PDV arrangement, (b) Two-frequency PDV arrangement.

magnification between images. Post-processing cannot, however, correct for the almost inevitable differences in the optical distortions or for the polarisation sensitivity of the beamsplitter. Even so-called polarisation-insensitive beamsplitters actually have reflectivity variations of up to about 5% for orthogonal linear polarisation states.

The technique described in this paper offers a solution to the pixel-matching problem. One or more acousto-optic modulators are used both to switch the laser beam and to alter the original frequency. The laser is tuned just off the low frequency side of the absorption line. This is used as the 'reference frequency'. The AOMs are chosen to produce a total frequency shift of about 600 MHz, sufficient to take the optical frequency from the original position off the side of the iodine absorption line to a 'signal frequency' close to the 50% absorption level on this side of the line. Thus, illumination beams appropriate for acquisition of both reference and signal images can be derived from a single source, and a common path geometry becomes possible, where the PDV system is reduced simply to an iodine cell positioned directly in front of the lens of the solid-state camera. The wavelength difference is negligible in comparison with the typical size of scattering particles, therefore any differences in scattering efficiency between the two beams are negligible. Exact alignment of the reference and signal images on the active area of the camera is automatic.

Precisely simultaneous image acquisition is clearly not possible. For steady-state flows this is not a major difficulty, provided that the seeding is relatively dense and the seeding distribution remains essentially unchanged between acquisition of the two image frames. The accuracy of measurements made in a steady-state flow can, of course, be improved by taking an average of the normalised intensity maps over an extended period.

2. Experimental Method

2.1 Illumination system

The arrangement of figure 2 was used to generate co-linear beams at two closely spaced optical frequencies for illumination of the single-camera PDV system. The components were mounted on a rigid 100 x 50 cm aluminium plate for portability. In the figure, solid lines represent light used to form the illumination sheet and dashed lines represent unused light. In figure 2(a), AOM#1 is switched on and AOM#2 is off. The beam experiences an 80 MHz downshift on passing through AOM#1 and is redirected towards the sheet-forming optics. In figure 2(b), AOM#1 is off and AOM#2 is on. The beam passes unchanged in frequency or direction through AOM#1 and then makes a double pass through AOM#2, experiencing an

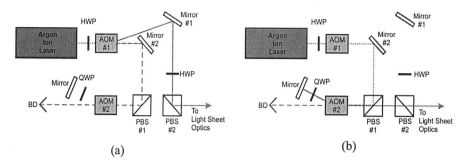

Figure 2. Schematic of the two-frequency PDV system showing: (a) reference beam path (b) signal beam path. HWP – half-wave plate, QWP – quarter-wave plate, AOM – acousto-optic modulator, PBS – polarising beam splitter, BD – beam dump.

upshift of 520 MHz. The quarter-wave plate in this arm rotates the plane of polarisation, allowing the beam to pass through PBS#1 and PBS#2 towards the sheet-forming optics. The half-wave plates in the system ensure correct routing of the beams.

Both output beams were coupled into a single-mode, polarisation-preserving, optical fibre [6] for transmission to the sheet-forming optics. This produces a clean Gaussian beam and ensures that the two light-sheets are exactly superimposed.

The source used was an argon-ion laser with an etalon, tuned to 514.5 nm. The single-mode linewidth was about 3 MHz. Although not designed to be tuneable, the etalon temperature could be altered using a potentiometer on the laser backplate. The setting resolution was about 70-100 MHz. The iodine cell itself was used to monitor frequency variations, since there was no optical frequency indicator on the laser.

The laser was tuned just off the iodine absorption line. The frequency of the upshifted beam then automatically lies at a position somewhere on the low-frequency side of the iodine absorption line. This beam was used to obtain the signal images in the PDV system. Both signal and reference beams experience scattering from the flow, and the frequency shifts imposed on each beam are identical. Since both beams also pass through the iodine cell, the initial reference frequency must be positioned sufficiently far from the top of the absorption line that the maximum Doppler upshift expected in the system will not bring it over the edge of the absorption line.

2.2 Image-capture system

The image-capture system is greatly simplified compared with that required for a single frequency of illumination. Both reference and signal beams pass through the iodine cell, and the system is reduced to a single solid state camera incorporating a zoom lens, together with a temperature-stabilised iodine cell positioned directly in front of the lens.

The camera used is a 1K x 1K Peltier-cooled 12-bit, digital Imager 3 supplied by LaVision. Dedicated 'DaVis' image acquisition and processing software is used for control and image-processing. The iodine cell, which operates in the saturated mode [7], is 25 mm in diameter and 50 mm long, with a cold finger held at 40° C and the cell body at 56.5° C.

Accurate velocity measurements using PDV require an exact knowledge of the shape of the iodine absorption function. This measurement was made independently using a Spectra-Physics Argon-ion laser with a highly stabilised etalon and frequency-scanning capability. The iodine transmission as a function of optical frequency was characterised using a sixth-order polynomial fit.

A simpler experimental arrangement including only AOM#2 was actually used to obtain the images presented in this paper. AOM#1 was unavailable at the time and, as a result, the reference beam was permanently 'on'. An additional subtraction step was included in the processing to separate the reference and signal images before taking the ratio.

Synchronisation of image capture with the appropriate illumination beams was achieved using a TTL signal from the camera to switch the AOM. In these experiments, 40 reference/signal image pairs were acquired and stored in memory in about 15 seconds, using a camera exposure time of 100 ms. The images were later saved in batches to hard disk.

The test object used to assess the PDV system was a rotating perspex disc 150 mm in diameter. The maximum circumferential velocity of the wheel was 31 ms^{-1}. The illumination direction was 5° from the plane of the disc and the observation direction 30° from the same plane; a near backscatter arrangement. The geometry is chosen to have high sensitivity to the horizontal component of rotational velocity in the plane of the wheel. For measurements on the disc, it was not necessary to form the diverging illumination beam from the fibre into a light sheet, since the front surface of the disc itself defined the measurement plane.

Background images were acquired by positioning a small circle of card close to the output end of the optical fibre, to throw the surface of the wheel alone into shadow while retaining any scattering from surrounding objects that would also be present during velocity measurements. Baffles were arranged to reduce unwanted scattering as much as possible. Background subtractions were performed routinely on all images subsequently acquired by the camera.

The expected frequency shift is given by the usual Doppler equation [8] $\Delta v = v(\overline{\mathbf{o}} - \overline{\mathbf{i}}).\mathbf{v}/c$ where v is the optical frequency, $\overline{\mathbf{o}}$ and $\overline{\mathbf{i}}$ are unit vectors in the observation and illumination directions respectively, $\mathbf{v}$ is the velocity vector and c is the free-space speed of light. A maximum frequency shift of +/- 104 MHz is expected at the highest and lowest points on the spinning disc, where the velocity vector is horizontal. The horizontal component of velocity should vary linearly along any vertical line through the disc.

3. Results

Sets of image pairs were stored for anticlockwise rotation of the disc, at the maximum angular velocity. Forty image pairs were acquired, the AOM being synchronised with the camera shutter to provide the appropriate illumination for each image. The images were saved to disc as pairs: image RS = reference + signal illumination and image R = reference illumination for each pair.

The intensity profile of the illumination beam results in high greyscale values for pixels near the centre of the disc, and lower values towards the edges of the disc. The initial image analysis, which is carried out using the DaVis 6.0 software by loading the stored images into memory buffers, is simply a subtraction S = RS - R, followed by a division to obtain an image N = S / R. The values in N should be multiplied by a constant to obtain the normalised intensity map. If the signal and reference illumination intensities are equal, the multiplication factor is 1. The intensity at the centre of the disc in each image corresponds to the unshifted laser beam frequency, since the velocity here is zero. Photodetector measurements of the signal and reference beam intensities gave a signal/reference intensity ratio of 1.17. This multiplication factor was used to correct the normalised images. The greyscale values at the centre of the disc then give the normalised iodine transmission for the unshifted laser frequency. The value was found to be 0.92 for our experiment. The normalised images are quite noisy therefore, when measuring the greyscale level at any position in an image, it was not sufficient to take the value for a single pixel. Instead, an average was taken over a region 10 pixels wide by 10 high centred on the position of interest.

For all pairs of images in the data series, the normalised intensity maps were calculated. Each map was corrected for the variations in signal beam intensity, the correction factor being chosen to produce the expected value for the zero-velocity absorption at the centre of the disc. An average was then taken of the entire set of 40 normalised images in the series. The resulting image is shown in figure 3(a). The difficulty in tuning the laser used to a predetermined frequency, has resulted in the unshifted frequency being quite close to the top of the iodine absorption curve, and thus the relationship of image intensity to Doppler shift in figure 3 is non-linear, as seen on the graph ordinate.

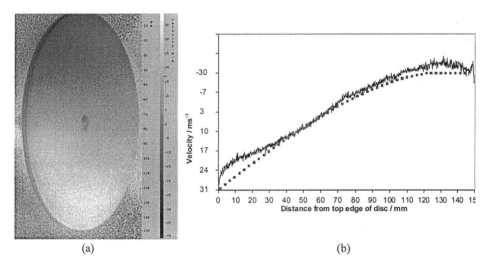

(a) (b)

Figure 3: (a) Normalised intensity map obtained by averaging over the entire set of 40 images for anticlockwise rotation; (b) Vertical profile through the image of (a), close to disc centre.

The performance of the PDV system can be assessed by comparison of the data with independent measurements of the disc velocity. A profile was taken vertically through the image of figure 3(a), 15 pixels to one side of the disc centre to avoid the specular reflection from the central spindle. Figure 3(b) shows a profile taken through the image of figure 3(a), together with the relevant section of the polynomial fit to the iodine transmission curve, calculated using independent tachometer measurements of disc rotation frequency. The iodine curve is positioned by selecting the transmission on the horizontal centre line of the disc to match the value determined experimentally.

The PDV measurements match the calculated transmission values very well over the central region of the disc. The maximum error in greyscale value in this region is about +/-8, corresponding to a velocity error of +/-1 ms^{-1}. Towards the edges of the image the fit is less satisfactory, and the experimental values are higher than those expected theoretically for both the top and bottom edges of the disc. The maximum disparity in greyscale level between the two curves is about 50, corresponding to a velocity error of 7 ms^{-1}. This is attributed to low illumination intensity in the outer regions of the disc. The greyscale levels in the original images fall to only about 100 towards the disc edges and, although background levels were subtracted from the images as described above, small variations are likely because of the changing signal illumination levels. Any systematic errors will be increased by the signal processing and will be worst in regions of low intensity.

4. Discussion

The results show that this technique can provide accurate PDV velocity data from a spinning disc. The accuracy and resolution of the information deteriorate in regions of low illumination intensity. This problem is, however, universal in PDV measurements and can be overcome by the use of a more powerful laser and perhaps the introduction of beam-shaping optics to produce a more uniform illumination beam. Processing is greatly simplified in the one camera system by the avoidance of any pixel-matching requirement, and errors caused by the polarisation sensitivity of the beam splitter in a conventional system are eliminated.

In these experiments, the optical power in the scattered beam remains constant with time, as it is always integrated over at least one complete rotation of the disc. In a real flow the scattered power, and its distribution over the camera image, may change between acquisition of the reference and signal images. In a steady-state flow, if the seeding remains fairly uniform, the errors should remain small and can be minimised by averaging over many normalised images. The degree of averaging required will depend on the nature of the flow.

In a rapidly time-varying flow, of course, averaging over many images is not an option, and application of the technique is not straightforward. A pulsed laser would probably be required for this type of flow, to freeze the motion of the particles [9]. With state-of-the-art cameras, the time between two frames can be as short as 100 ns, enabling near-simultaneous acquisition. This would be acceptable for many moderate-speed flows.

5. Conclusions

Two closely-spaced optical frequencies, produced by switching an acousto-optic modulator, can form the basis of a common-path, single-camera PDV system. Superposition of the signal and reference images is achieved automatically, albeit at the expense of precise simultaneity in the acquisition of the image pair. The performance of the system has been assessed using a rotating disc as a test flow and a velocity resolution of +/-1 ms^{-1} has been achieved. The potential benefit of the technique is a major simplification of the PDV image-capture system for time-invariant flows

References

[1] Komine H, Brosnan S, Litton, A and Stappaerts E 1991 *AIAA Paper 91-0337*
[2] Meyers J F 1995 *Meas. Sci. Technol.* **6** 769-783
[3] ChanV S S, Heyes A L, Robinson D I and Turner J T 1995 *Meas. Sci. Technol.* **6** 784-794
[4] Ford H D and Tatam R P 1997 *Optics and Lasers in Engineering* **27** 675-696
[5] Manners R J, Thorpe S J and Ainsworth R W 1996 *Proc. I. Mech. E. Conference on Optical Methods and Data Processing in Heat and Fluid Flow* (London)
[6] Rashleigh S C 1983 *Journal of Lightwave Technology* **LT-1** 312-331
[7] Quinn T J and Chartier J-M *IEEE Transactions on Instrumentation and Meas.* 1993 **42** 405-406
[8] Irani E and Miller L S 1995 *Society of Automotive Engineers, Paper 951427*
[9] McKenzie R L 1995 *AIAA Paper 95-0297*

Inst. Phys. Conf. Ser. No. 177
Paper presented at 1st Int. Conf. on Optical & Laser Diagnostics, London, 16–20 Dec. 2002

An Investigation of Underexpanded Free Jets from Straight Nozzles

L N Ung, G K Hargrave

Wolfson School, Loughborough University, Loughborough, LE11 3TU, UK

Abstract. Good quantitative 2-dimensional experimental measurements of the near field structures of underexpanded free jets are required for the accurate validation of CFD codes that predicts gas dispersion and safety zones during venting, flaring or accidental gas releases from high pressure pipelines. Underexpanded jet is formed in gas releases when the pressure ratio of exit pressure to atmospheric pressure is greater than 1.89. In this paper, velocity measurements of highly underexpanded jet issuing from a nozzle of 3mm diameter, length-to-diameter ratio of 5 is investigated using particle image velocimetry. Here, the plenum gauge pressure investigated is 3.5 bar.

1. Introduction

The safe design of installations incorporating high pressure pipework and gas handling plant requires the engineering of adequate relief systems. Venting and flaring of natural gas is commonly practiced in order to achieve pressure relief from high pressure handling plant or pipework. The safety of the operation is of utmost importance. The mechanism of gas dispersion into the atmosphere during gas venting is highly dependent on the way in which the gas is released at source. Depending on the initial velocity of the release, the dispersion processes are principally governed by the momentum and density together with the prevailing atmospheric conditions. Hence, there is interest in the effect of jet source structure on the subsequent plume dispersion in the far field.

In gas releases having a pressure ratio of exit pressure to atmospheric pressure greater than 1.89, underexpanded jet will be formed. A moderately underexpanded jet will decay to a subsonic flow through a series of expansion and compression waves forming shock cells downstream of the jet, as illustrated in Figure 1.

As the pressure ratio is increased further, the jet will become highly underexpanded and a Mach disk will be formed. A Mach disk is a thin layer of a strong normal shock. Downstream of the Mach disk, shock cells will be formed, and the jet decay to that of a subsonic flow. A highly underexpanded jet is illustrated in Figure 2. Flow just before the Mach disk is supersonic with Mach number, $M \gg 1$. The flow immediately after the disk will become subsonic ($M < 1$).

40

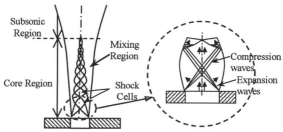

Figure 1: Moderately underexpanded jets.

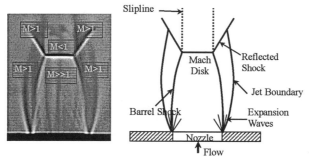

Figure 2: Highly underexpanded jet.

Currently the near field structure of underexpanded jets are not widely studied. Literature is available for qualitative measurements of underexpanded jets, but there is a paucity of quantitative experimental data on the near field structure of jets with high pressure ratios. Quantitative data for underexpanded jets available are that of a far field jet downstream from the nozzle exit [1]. The near field data are vital for the validation of CFD codes to predict the plume dispersion in the far field and hence, the prediction of safety zones during gas venting or accidental gas releases at high pressure handling plant. These gas releases can be hazardous, especially when the released gas is toxic or flammable. The ability to predict the gas dispersion from a high pressure jet is desirable for risk assessment and hazard management.

Previously, CFD was used to predict the structure of underexpanded jets which was validated using far field data. with the assumption of a psuedo-source[2]. Data from either schlieren imaging or shadowgraphy providing Mach disk diameter and barrel shock [3], laser-doppler velocimetry measuring velocity at a point [4], or total pressure measurements obtained using intrusive methods [5][6] were also used for the validation. Several researchers have investigated the use of other techniques such as particle image velocimetry [7][8], planar doppler velocimetry [9] and Rayleigh scattering [10], for high speed flow application.

The aim of this study is to provide detailed velocity maps of the near field structure of highly underexpanded jets. Data will be presented from shadowgraphy imaging of the flow structure together with 2-dimensional velocity maps obtained using digital, cross-correlation Particle Image Velocimetry (PIV). The long term aim of the project is to investigate the structures of the jet issuing from various nozzle dimensions.

2. Method

Figure 3 shows the schematic of the experimental set-up for PIV. The nozzle dimension chosen for this investigation was a straight nozzle with diameter 3mm and length-to-diameter ratio of 5. The light source used was a double-pulsed Nd:Yag laser operating at 532nm with 8ns pulse width and a pulse energy of 50 mJ. The laser light was formed into a sheet 40 mm high and 1 mm thick and illuminated the region immediately downstream of the exit nozzle. The particle images were captured by a Kodak ES1.0 two-frame CCD camera with a resolution of 1000 X 1016 pixels placed perpendicular to the plane of the laser sheet. The laser and camera system were operated to provide a particle image separation of 0.7 µs. Cross-correlation analysis for the derivation of the two-dimensional velocity field was achieved using TSI Insight 3.0 software. The interrogation region was 64 by 64 pixels with 75% overlap. The image resolution is 12.69 µm/pixel.

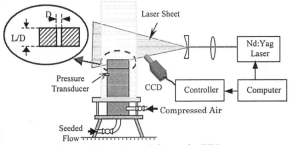

Figure 3: Experimental Set-up for PIV

Seeding was introduced to both the jet and the ambient region of the jet. The seeding used was 1 micron white alumina particles. The seeding used for both regions was the same and seeding levels were controlled to maintain a constant seeding density in order to reduce velocity bias.

3. Results and Discussion

Figure 4 shows the shadowgraph of underexpanded jets from the straight nozzle of 3mm diameter and length to diameter ratio of 5 at the respective plenum gauge pressure. These shadowgraphs cover the region from the nozzle exit to 4 diameters downstream. At 2.0 bar, the shadowgraph shows a moderately underexpanded jet. Highly underexpanded jet can be seen by the shadowgraph from pressure of 3.5 bar. The features of a underexpanded jet can be clearly seen from these shadowgraphs. In a moderately underexpanded jet, shock diamonds is formed by the compression and expansion waves. The main features of a highly underexpanded jet consists of a Mach disk, barrel shock, slipline, and expansion waves and can be clearly observed from the shadowgraphs. As the plenum pressure increases, the Mach disk diameter and the barrel length (distance from the nozzle exit to the Mach disk) will increase. This phenomena was previously investigated. At plenum gauge pressure of 3.5 bar, the Mack disk diameter and barrel length measured from the shadowgraphs are 3.11mm and 0.34mm respectively. The following results from PIV is that of an average from 250 image pairs.

42

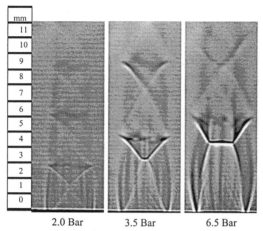

2.0 Bar   3.5 Bar   6.5 Bar

Figure 4: Shadowgraphs of jet issuing from a 3mm, L/D=5 nozzle at 2.0, 3.5 and 6.5 bar gauge plenum pressure respectively.

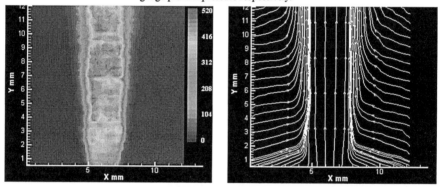

Figure 5: Velocity contour plot and streamline plot of a jet from a 3mm diameter, L/D = 5 nozzle at 3.5 bar plenum gauge pressure

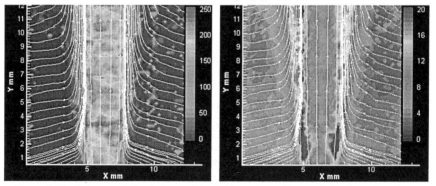

Figure 6: V turbulence and U turbulence of the jet from a 3mm diameter, L/D=5 nozzle at 3.5 bar plenum gauge pressure.

Figure 5 shows the velocity map and the streamline of the jet at 3.5 bar plenum gauge pressure. The area of view covers the region from the nozzle exit to 4 diameters downstream. The center of the jet is at 6.2mm on the X scale in the plots. It can be observed that the highest velocity of the jet is at the center of the jet. From the streamlines, entrainment into the jet plume can be observed. From the velocity map, there is a thin region of low velocity at approximately 3.5mm from the nozzle exit, as compared to the velocity across it.

Figure 6 shows the V-turbulence and U-turbulence plots of the jet with streamlines. From the plots, it can be seen that the turbulence in the v-direction (upwards) is high near the Mach disk at about 3.5mm from the nozzle exit. From the u-turbulence plots, it can be seen that the u-turbulence is the highest at the region between the jet boundary and the barrel shock where compression waves exist. The v-turbulence is much higher than the u-turbulence. The u and v-turbulence is relatively high at the shear layer of the flow. Turbulence in the ambient region is relatively low.

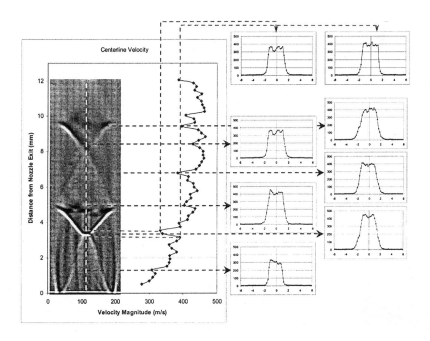

Figure 7: Axial and transverse velocity plots of the jet with comparison with shadowgraph.

Figure 7 shows the both axial and transverse velocity plots of the jet. The axial plot is the centerline velocity of the jet. Velocity plots at different positions from the nozzle exit are plotted in the transverse direction. These plots correspond to the positions on the shadowgraph and at various points where there's a dip in velocity at the centerline.

It can be seen that velocity is highest in the center of the jet. Across the Mach disk, the velocity decreases. The velocity increases from the exit to the Mach disk. Comparing the plots, velocity of the jet plume at the Mach disk region is the highest.

The position of the Mach disk cannot be clearly seen from the PIV results. This may be because the Mach disk is small. The diameter of the Mach disk diameter for the current PIV investigation is 0.34mm. In this case, each interrogation region used is 0.81mm. The size of the interrogation region was calculated based on the maximum velocity expected in the jet plume. The size of the image required was determined from the size of the jet and the ability to measure the flow at the ambient. From the results, it can be seen that there is a wide range of velocity magnitudes to measure in underexpanded jets. Hence, acquiring measurements of the complete plane in one realisation in underexpanded jet requires either a PIV system with very large dynamic range, or measurements that include variation in temporal displacement of the PIV images to focus on the different regions of the flow field.

4. Conclusions

This paper has described an investigation of the application of Particle Image Velocimetry to the study of highly underexpanded jets. Two-dimensional velocity data has shown structure coincident with the shadowgraph images and indicated velocity magnitudes consistent with basic flow theory for underexpanded jets. It is clear from this study that control of the size and location of the PIV interrogation region is essential in defining the velocity field in the regions of high velocity gradient if errors are to be minimised. Also, due to the very large dynamic range to be measured (from jet flow to entrainment flow), variation in temporal displacement of the PIV image pair, to focus on the different regions of the flow field, is essential for future investigations of underexpanded jets.

References

[1] Ewan, B C R and Moodie K 1986 *Comb. Sci. and Tech.* **45** 275-288
[2] Birch A D, Hughes D J and Swaffield F 1987 *Comb. Sci and Tech.* **52** 161-171
[3] Crist S, Sherman P M and Glass D R 1966 *AIAA J.* **4** 68-71
[4] Eggins P L and Jackson D A 1974 *J. Phys. D: Appl. Phys.* **7** 1894-1906
[5] Donaldson C D and Snedeker R S 1971 *J. Fluid Mechanics* **45** 281-319
[6] Seiner J M and Norum T D 1979 *AIAA Paper* No. 79-1526.
[7] Lawson N J, Page G J, Halliwell N A and Coupland J M 1999 *AIAA J.* **37** 798-804
[8] Krothapalli A, Wishart D P and Lourenco L M 1994 Proc. 7th Int. Symp. on Applications of Laser Techniques to Fluid Mechanics
[9] Samimy M and Wernet M P 2000 *AIAA J.* **38** 553-574
[10] Panda J and Seasholtz R G 1999 *Phys. of Fluids* **11** 3761-3777

Inst. Phys. Conf. Ser. No. 177
Paper presented at 1st Int. Conf. on Optical & Laser Diagnostics, London, 16–20 Dec. 2002
©2003 IOP Publishing Ltd

Instantaneous three-dimensional visualization of concentration distributions in turbulent flows with a single laser

A Hoffmann, F Zimmermann, C Schulz

Physikalisch-Chemisches Institut, University of Heidelberg, Im Neuenheimer Feld 253, 69120 Heidelberg, Germany

Abstract. A new laser-based technique for measuring instantaneous three-dimensional species concentration distributions in turbulent flows is presented. The laser beam from a single laser is formed into two crossed light sheets that illuminate the area of interest. The signal light from both planes is detected with a single camera via a mirror arrangement. Image processing enables the reconstruction of the three-dimensional data set in close proximity to the cutting line of the two light sheets. Volume visualization by digital image processing gives unique insight into the three-dimensional structures within the turbulent processes. Three-dimensional intensity gradients are computed and compared to the two-dimensional projections obtained from the two directly observed planes.

We apply this technique to measurements of the hydroxyl (OH) concentration distribution by laser-induced fluorescence (LIF) in a turbulent methane-air flame upon excitation at 248 nm with a tunable KrF excimer laser. Further measurements address the three-dimensional distribution of toluene-LIF in a turbulent, non-reactive mixing process of toluene and air.

1. Introduction

Understanding the three-dimensional structure of turbulent non-reactive and reactive flows is of major interest for the modeling of turbulent combustion processes which are omnipresent in practical combustion apparatus. In recent years laser diagnostics have proved to give insight in turbulent processes with high spatial and temporal resolution [1,2]. Statistical information from point measurements (i.e. Raman scattering and laser-induced fluorescence (LIF) [3]) has been an important input to flame simulations based on probability density functions (PDF) [4]. Two-dimensional imaging of LIF of many combustion relevant species and detection of Rayleigh and Raman scattering yielded quantitative scalar information on species concentration and temperature. This has been used for comparison with results of computational direct numerical simulations. These images also show curvature of flame fronts and concentration and temperature gradients. The latter being of major importance for statistical PDF simulations. However, these measurements only show cross-sections of the investigated object. The structural information and the gradients are distorted by the 2d-projection and no information is available how or if observed structures are connected with each other via the third dimension.

Several attempts have been presented to include the third spatial component into the detection system and to obtain three-dimensional information at least with limited spatial resolution by simultaneous or quasi-simultaneous observation of parallel or crossing planes.

Sweeping long-pulse or cw laser beams via scanning mirrors through the object was suggested by Yip et al. [5]. Recent approaches with multi-laser excitation of 3 [6] and up to 8 planes [7] with subsequent imaging with multiple cameras or a high-speed camera yielded new results, however, at high experimental costs. Further approaches used the different harmonics from a Nd:YAG laser to simultaneously image four adjacent planes on one camera with a filtered 4-faceted multi-imaging element [8]. In many cases, however, the parallel light-sheets are too far from each other to provide the spatial information necessary to resolve even small turbulent structures and to calculate gradients within the observed 3-dimensional data field.

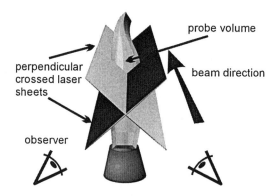

Fig. 1: Arrangement of the crossed laser light sheets.

Another method is therefore based on the observation of signal from two perpendicularly crossing planes (figure 1). This approach has been used by Knaus et al. for observations of the flame front orientation in an internal combustion engine [9]. They applied excitation with a quadrupled Nd:YAG laser and detection with two cameras. The spatial resolution of this technique is high along the crossing line but decreases rapidly with the increasing distance between the planes. It is therefore suited to yield three-dimensional gradients along the line of the crossing light sheets. In this work we follow this attempt using a single laser and camera only. Digital image processing is used to evaluate signal intensity gradients along the cutting line of the observed planes. Furthermore, a 3d-reconstruction of the volume around the cutting line is used to visualize the turbulent field.

2. Experiment

The rectangular laser beam profile of a tunable KrF excimer laser at 248 nm is reduced to a homogeneous, circular profile with a circular aperture and split with a dielectric beam splitter into two beams with equal intensities. Both beams are directed into the probe volume and focused with perpendicularly oriented cylindrical lenses forming an illumination pattern according to the sketch in figure 1. The laser-induced signal from both planes is directed to a spherical mirror via a combination of two aluminum coated mirrors for each plane and then focused by the aluminum coated spherical mirror (250 mm diameter, 350 mm focal length) towards two adjustable folding mirrors (18 mm diameter). They finally direct the two partial images onto the 25 mm image intensifier of a *LaVision* Flamestar 3 ICCD camera (figure 2). This allows to arrange the two images without overlap as shown in the small screen shot in figure 2. The size of these mirrors is limited by two effects. First, large mirrors would block a significant portion of the incoming light; second, larger mirrors would result in blurring of the images. The latter is due to the fact that the imaging quality of the spherical mirror is decreasing the farther away the object or image gets from its optical axis.

Elastically scattered light was suppressed by a dielectric mirror (248 nm, 0°, fwhm 15 nm, *Laser Optik*) in front of the camera.

This setup was used to either excite OH in a turbulent Bunsen type flame with excitation on the A-X(3,0) $P_2(8)$ line at 248.46 nm or to observe the mixing between air and a nitrogen stream saturated with toluene with the signal representing the resulting toluene / air ratio [10] using the same excitation wavelength.

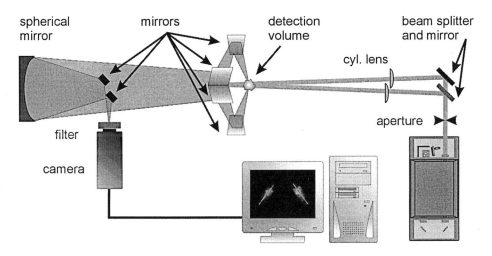

Fig. 2: Experimental setup of the single-laser single-camera measurements.

3.   Data evaluation and results

Different steps in image post processing are required in order to reconstruct the three-dimensional data matrix. First, the two partial images must be separated from the image the camera sees and must be mapped onto each other, second, the signal from the respective other plane must be removed from the partial images, third, the missing data between the observed plane must be interpolated, fourth, along the intersection line the three-dimensional gradients must be calculated from the 3d-data matrix and fifth, the complete 3d matrix must be visualized.

The two partial images are separated and mapped onto each other using additional images of a test object showing a rectangular grid. Since both partial images are well separated on the chip, it is sufficient to cut them out and then adjust them to the standard grid with a multi-parameter mapping function offered by the software used for camera control (Davis by *LaVision*).

As a consequence of the simultaneous illumination and detection of both planes a line of strong signal is detected in the middle of each partial image (upper frames in fig. 3). This signal is the projection of the second onto the first plane and vice versa. Its signal intensity, however, corresponds to the intensity measured in the respective other image after summation perpendicular to the beam direction. In a first step, this signal intensity is subtracted from the images. This procedure, however, still leaves some signal variation in the close proximity of the intersecting line. Therefore signal in this area is removed (four pixel wide) and then refilled via non-linear diffusion as explained in the following paragraph. This procedure is not fully satisfactory since information is removed from the region of highest interest, which causes a reduction in spatial resolution. Further experiments, which are currently in progress, solve this problem by temporally delaying one of the two laser beams and by separately detecting the signal from both planes with two independent cameras.

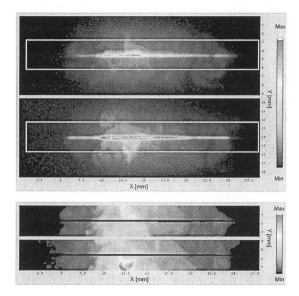

Fig. 3: Image pair from the two perpendicular planes in the toluene seeded nitrogen / air mixing flow. Upper frames: Raw images correctly aligned and mapped, lower frames: Same images reduced to a smaller size after subtraction of the unwanted signal contributions from the other light sheet.

The obtained two planar data sets are then written into a three-dimensional matrix according to their relative orientation in the probe volume. A non-linear diffusion algorithm [11] is then applied to the 3d-data matrix distributing the information from the two planes into the empty space in between. The original data in the measured planes is kept constant by setting them to their original values after each diffusion step. These steps fill the whole matrix including the central gap (see fig. 3, lower frames) with interpolated data. The quality of the interpolated data decreases with distance to the planes with original measured data. From the three-dimensional data set cross sections in any direction can be depicted. The images in figure 4 show some cross sections of the calculated volumetric data obtained from toluene-LIF measurements in a turbulent mixing process. The first image pair represents the actually measured planes, the second and third image pair shows interpolated planes 2 and 8 pixels away from the light-sheet illuminated planes (1 pixel equals ~70 µm). As an inherent consequence of the diffusion process the interpolated data become diffuse with increasing distance from the source planes. This is most obvious in the lower, square plots in figure 4 that show cuts through the reconstructed data matrix perpendicular to both light sheets with the original data in the vertical and horizontal axis. However, for many purposes in understanding turbulent processes and in modeling turbulent flows the obtained three-dimensional information up to several millimeters around the cutting line is sufficient to depict the turbulent structures and to determine gradients in species distributions. Nonetheless, we have to notice that in the cutting line we obtain real three-dimensional data. For each point on this line we know its relative value and all three components of its spatial variation.

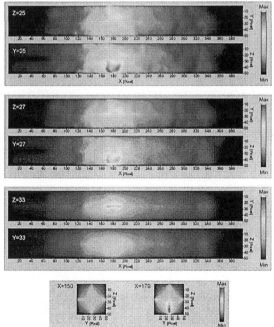

Fig. 4: Cross-sections through the 3d-data matrix for the toluene seeded nitrogen / air mixing process. First image pair: Originally measured planes (the gap seen in Fig. 3 is now filled by the diffusion process), middle and lower image pair at 2 and 8 pixel distance respectively. Bottom frames: Cross sections perpendicular to both original planes at two different x positions.

The resulting three-dimensional data matrix can also be visualized by a 3d-visualization tool that assigns not only color but also transparency values to the given intensity values [12]. The results for the toluene mixing shown in figures 3 and 4 and the OH measurements in the lean Bunsen flame is presented in fig. 5. The flame image shows a measurements position in the lower portion of the Bunsen flame. In the middle of the image no signal is detected in the incoming, unburned fresh gases, whereas the dark areas at both sides of the image represent the limit of the flame towards the surrounding air. The front side of the visualized volume elements is one of the measured planes, the depth of the visualized 3d-volume is 2 mm.

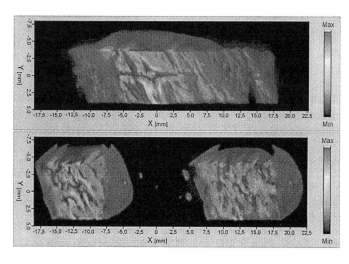

Fig. 5: Volume visualization of the toluene-LIF distribution in a turbulent mixing process and the OH-LIF distribution in a turbulent lean Bunsen flame.

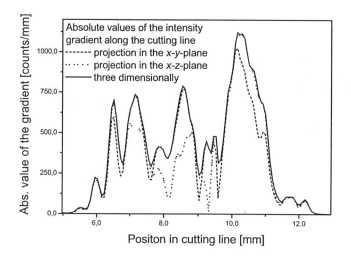

Fig. 6: Comparison of absolute values of intensity gradients obtained from both measured planes and from the 3d-reconstructed data matrix (solid line).

Figure 6 finally shows the absolute values of intensity gradients for each pixel along the cutting line of the two observed planes in the lean Bunsen flame (left portion of the image shown in figure 5). It is obvious that the projection of the intensity gradients which are observed in either 2d-image might deviate significantly from the "real" 3d-gradient depending on the orientation of the gradient relative to the observed plane. A correlation analysis of local gradients vs. local concentrations as desired for PDF analysis therefore requires the full three-dimensional information.

4. Conclusions

A new approach to instantaneously measure three-dimensional LIF-intensity distributions in turbulent systems is presented in toluene-seeded nitrogen/air mixing processes and in a turbulent lean Bunsen flame. The method yields high spatial resolution along the crossing line of two simultaneously observed perpendicular planes. It is therefore ideal to measure three-dimensional concentration gradients along this intersection line. In the close vicinity of the crossing line non-linear diffusion allows the reconstruction of the 3d-data matrix that then can be visualized by appropriate volume visualization tools.

The attempt presented here requires a single laser and camera only. However, this approach requires interpolation of the data in the close proximity of the cutting line and it is limited by the poor optical throughput of the multiple-mirror arrangement. Further measurements using the same image post-processing with temporally delayed illumination (via an optical delay line) and separate observation of both planes with two cameras will solve these problems. These measurements are currently under way.

References

[1]   Wolfrum J 1998 *Proc. Combust. Inst.* **27** 1
[2]   Böckle S, Kazenwadel J, Kunzelmann T, Shin D-I, Schulz C, Wolfrum J 2000 *Proc. Combust. Inst.* **28** 279
[3]   Masri A R, Dibble R W, Barlow R S 1996 *Prog. Energy Combust. Sci.* **22** 307
[4]   Nau M, Wölfert A, Maas U, Warnatz J 1995 8th *Int. Symp. on Transport Phenomena in Combustion,* 605
[5]   Yip B, Fourguette D C, Long M B 1986 *Appl. Opt.* **25** 3919
[6]   Landenfeld T, Kremer A, Hassel E P, Janicka J, Schäfer T, Kazenwadel J, Schulz C, Wolfrum J 1998 *Proc. Combust. Inst.* **27** 1023
[7]   Nygren J, Hult J, Richter M, Aldén M, Christensen M, Hultqvist A and Johansson B 2002 *Proc. Combust. Inst.* **29** in press
[8]   Frank J H, Lyons K M, Long M B 1991 *Opt. Lett.* **16** (12) 958
[9]   Knaus D A, Gouldin F C, Hinze P C, Miles P C 1999 *SAE technical paper* 1999-01-3543
[10]   Koban W, Schorr J, Schulz C 2001 *Appl. Phys. B* **74** 111
[11]   Scharr H, Jähne B, Böckle S, Kazenwadel J, Kunzelmann T, Schulz C 2000 in: Sommer G, Krüger N, Perwass C (Eds.), Mustererkennung 2000, 22. DAGM Symposium, Kiel 2000, *Informatik aktuell, Springer Verlag, Heidelberg* 325
[12]   Dartu C 2001 Dissertation, IWR, Universität Heidelberg

Inst. Phys. Conf. Ser. No. 177
Paper presented at 1st Int. Conf. on Optical & Laser Diagnostics, London, 16–20 Dec. 2002

Characterisation of drop impact on heated surfaces by optical techniques

G E Cossali, M Marengo, M Santini

Università di Bergamo- Facoltà di Ingegneria- 24044 Dalmine (BG)-Italy

Abstract. The impact of liquid drops on heated surfaces is a phenomenon of relevance in many applied fields, like in direct injection internal combustion engine, where the fuel drops impact on the hot piston wall, or in spray cooling, etc. The first step to a deeper understanding of the phenomenon is to try to investigate the effects produced by the impact of single drops on a heated surface, among these the so called secondary atomisation, i.e. the formation of small droplets after the impact of the primary drop on the wall, is considered of paramount importance for many applications. The use of non-intrusive optical techniques is necessary in this field and the paper describes the application of three of such techniques to the analysis of the secondary atomisation produced by single millimetric drop impacting on a heated aluminium surface under different regimes of phase transition. Phase Doppler anemometry is used to measure the size and velocity of the secondary droplets in a size range between 2µm and 250µm. However, the impact produces droplets having also drops of larger diameter (up to 700 µm) and to extend the range of the measured size an image analysis technique based on a high resolution CCD camera was also used. The present configuration allowed to detect droplets having a size from 30µm on, and a proper analysis of the results from this technique, together with those coming from the PDA measurements, consented to obtain the "extended" secondary droplet size distribution from 2µm to few mm. A PIV (particle image velocimetry) technique was also applied to collect information about the velocity of the secondary droplets acquiring the images produced by a double shot strobolamp and using the cross-correlation method. Results are compared and the advantages and disadvantages of each technique are discussed.

Inst. Phys. Conf. Ser. No. 177
Paper presented at 1st Int. Conf. on Optical & Laser Diagnostics, London, 16–20 Dec. 2002
©2003 IOP Publishing Ltd

Time-Resolved Particle Image Velocimetry: Analysis Of Random Errors

K Anandarajah, G K Hargrave and N A Halliwell

Wolfson School of Mechanical and Manufacturing Engineering, Loughborough University, Loughborough, Leicestershire, LE11 3TU, United Kingdom.

Abstract. This paper describes a theoretical model of Particle Image Velocimetry, which is aimed at quantifying the random errors in a time-resolved PIV experiment. An estimation of these inherent errors is necessary before comparisons can be made with CFD code predictions. In particular, the importance of normalising the cross-correlation function, before curve fitting, to provide sub-pixel displacement measurements is examined.

1. Introduction

Particle Image Velocimetry (PIV) is now a well-established, non-intrusive measurement technique which provides instantaneous snapshots of velocity field maps, over a two-dimensional region within a flow. Advances in pulsed laser technology, digital image processing and data storage are now allowing PIV to be used in a time-resolved mode where velocity maps can be produced at kHz rates. This important advance makes PIV much more powerful as a tool for the refinement of CFD code predictions. It is now important, however, that the inherent random errors associated with time-resolved PIV measurements are quantified before comparisons are attempted.

One of the major challenges in PIV is increasing the measurement precision. Of particular interest are the errors associated with the post-processing of seeding particle images ranging from digitisation of the images to the location of the correlation peak centre. PIV accuracy has been enhanced by the introduction of sub-pixel interpolation where, typically, a Gaussian profile is fitted to the correlation peak and the centre estimated to an uncertainty of about 0.1 pixels. Inherent in a PIV measurement is a random error which occurs due to the random positioning of seeding particle images within a chosen interrogation region. In a time-resolved PIV experiment subsequent snapshots of the same interrogation region will produce changes in the measured velocity over time (which are of interest) and over space. The latter is due to the fact that at each snapshot the particle images are in a different spatial position within the interrogation region. In this way we need to establish the magnitude of the rms variation in velocity due to particle image position, which is a random error, before we can have confidence in deductions made from time varying statistics due to turbulent or unsteady flow. The rms variation in velocity due to particle image position is increased when velocity gradients exist across the interrogation region as described by Adrian (1997).

In this paper we model a time-resolved PIV experiment in order to quantify the spatial random error *in the absence* of velocity gradients across the interrogation region. We are focused on estimating the random error, which occurs from curve fitting to the correlation peak in order to obtain sub-pixel displacement measurement. The error accrues from minor changes in the correlation peak shape, in each subsequent measurement, in time-resolved PIV for the same interrogation region.

Huang et al. (1997) estimated the same errors, in a model experiment, by comparing the velocities measured from different interrogation regions for the case of a uniform particle image displacement across the full field of view. It is important to note that the noise characteristics in the experimental model used in Huang's case are different. The mean bias and rms were derived from many overlapping interrogation regions in a single snapshot of the flow field. Importantly, we have chosen to model the same interrogation region at different times to simulate a true time-resolved PIV experiment. Huang et al. (1997) noted a significant reduction in sub-pixel errors when the interrogation region images were normalised before cross-correlation. Towers and Towers (2001) also reported an increase in sub-pixel displacement error estimation when FFT based correlation techniques without normalisation. We have examined this effect and report the results. The effect on the random errors associated with non-clipped and clipped particles was also examined in this paper. The errors based on clipped particles are essential, because this issue closely resembles a 'real' PIV interrogation region in which, the presence of clipped particles at the boundary of the interrogation region do exist.

2.   The Experimental Model

The diffraction limited image of a seeding particle is an Airy disc which mathematically is defined by the square of a first order Bessel function shown in Goodman (1996). It is common in modelling a PIV experiment, however, Hart (1999) and Huang et al. (1997) use a two-dimensional Gaussian intensity profile, $I(x, y)$ which is a good approximation, and given by:

$$I(x,y) = I_o \, exp\left[\frac{-(x-x_o)^2 - (y-y_o)^2}{d_i^2}\right] \qquad (1)$$

The centre of the particle image is located at (x_o, y_o) with a peak intensity of I_o. The particle image diameter d_i is defined by the $e^{-1/2}$ intensity value of the Gaussian function, which by definition contains 61% of the total intensity in the envelope. The simulated two and three-dimensional expanded intensity profile images are shown in Figure 1(a) and (b) respectively. For comparison with Huang et al. (1997) results on an interrogation region size of 21 x 21 pixels were chosen and the particle image number density in this region was 22. The model distributes the Gaussian profile centre coordinates (x_{oi}, y_{oi}) randomly in the region as shown in Figure 2(a) and Figure 3(a) for both non-clipped and clipped particle images with a chosen image diameter of 2.8 pixels. With clipped particle images, the presence of whole particle for both exposure image pairs will reduce with increasing displacement. In order to examine the errors in 'real' PIV experiments, the particle images were digitised based on an 8-bit imaging device to produce a pixelated interrogation region image $I(x_p, y_p)$ the intensity at a pixel centre (x_p, y_p) for a pixel width (w) is calculated from:

$$I(x_p, y_p) = I_o \int_{x_p - \frac{w}{2}}^{x_p + \frac{w}{2}} \int_{y_p - \frac{w}{2}}^{y_p + \frac{w}{2}} exp\left[\frac{-[(x-x_o)^2 + (y-y_o)^2]}{d_i^2}\right] dxdy \qquad (2)$$

Figure 1(a) 2D Gaussian Profile

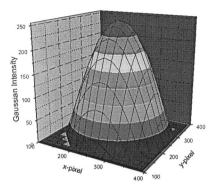

Figure 1(b) 3D Gaussian Profile

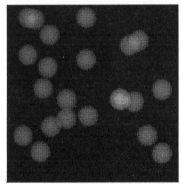

Figure 2(a) Interrogation region image

Figure 2(b) Pixelated interrogation region

Figure 2 Non-Clipped Particle Images

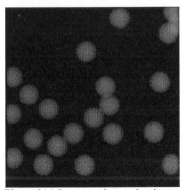

Figure 3(a) Interrogation region image

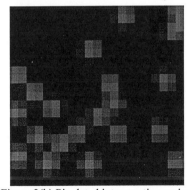

Figure 3(b) Pixelated interrogation region

Figure 3 Clipped Particle Images

A corresponding simulated Gaussian profile interrogation region image and pixelated interrogation region image for both non-clipped and clipped particle images is shown in Figure 2 and 3. Having established the first interrogation region image $I_1\ (x_p,\ y_p)$, the second image $I_2\ (x_p,\ y_p)$ was calculated where each particle image was shifted by one tenth of a pixel width $w/10$ in the x direction. This process was repeated to produce pairs of interrogation region images for particle image displacements of $nw/10$ where $n = 0,\ 1,\ 2, \ldots 10$.

3. Data Processing

We now proceed to compute the cross-correlation of the generated interrogation region images. The discrete cross-correlation function $R(m,n)$ of the two images $I_1(i,j)$ and $I_2(i,j)$ is given by:

$$R(m,n) = \sum_i \sum_j I_1(i,j) I_2(i-m, j-n) \tag{3}$$

where for convenience the pixel centre coordinates (x_p, y_p) now correspond to (i,j).

The correlation function $R(m,n)$ can be computed directly in the spatial domain or by a Fast Fourier Transform algorithm (FFT) in the frequency domain. The latter is extremely fast and efficient and is used in most commercially available PIV software analysis routines. Although simple to implement equation (3) has the disadvantage that it is sensitive to absolute intensity changes in I_1 and I_2 which may occur in experimental practice. This can be avoided by calculating a normalised discrete cross-correlation function $R_N(m,n)$ defined by

$$R_N(m,n) = \frac{\left\{ \sum_i \sum_j [I_1(i,j) - \bar{I}_1] \times [I_2(i-m, j-n) - \bar{I}_2] \right\}}{\left\{ \sum_i \sum_j [I_1(i,j) - \bar{I}_1]^2 \times \sum_i \sum_j [I_2(i,j) - \bar{I}_2]^2 \right\}^{1/2}} \tag{4}$$

where $\bar{I}_1, \bar{I}_2$ are spatial means of I_1, I_2 respectively. $\bar{I}_1, \bar{I}_2$ is computed and subtracted from the individual intensity values if these are greater than the average intensity. If the average intensities are smaller than the spatial means, they are set to zero as analysed by Pust (2000).

$R(m,n)$ and $R_N(m,n)$ are calculated directly in the spatial domain and an example is shown in Figure 4(a) and Figure 4(b) respectively.

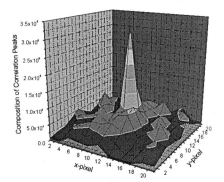
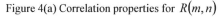
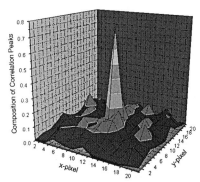

Figure 4(a) Correlation properties for $R(m,n)$ Figure 4(b) Correlation properties for $R_N(m,n)$

4. Correlation Peak Location

The error inherent in calculating the position of the centre of the correlation peaks defined by either equation (3) or (4) is ±0.5 pixels and it is necessary to locate the correlation peak more precisely by fitting a profile to the pixelated peak which is generated. It is common to fit a Gaussian profile due to the nature of the Gaussian correlation peak generated in the model experiment. If the centre of the pixelated correlation peak is given by (x_c, y_c), the actual centre (x_{pk}, y_{pk}) is estimated by Willert and Gharib (1991) to be situated at

$$x_{pk} = x_c + \frac{\{\log R(x_c - 1, y_c) - \log R(x_c + 1, y_c)\}}{2\{\log R(x_c - 1, y_c) + \log R(x_c + 1, y_c) - 2\log R(x_c, y_c)\}}$$

$$y_{pk} = y_c + \frac{\{\log R(x_c, y_c - 1) - \log R(x_c, y_c + 1)\}}{2\{\log R(x_c, y_c - 1) + \log R(x_c, y_c + 1) - 2\log R(x_c, y_c)\}} \quad (5)$$

The displacement of the correlation peak $\left(x_{pk}, y_{pk}\right)$ from the origin corresponding to the particle image displacement was calculated for each sub-pixel displacement as described in Section 2.

5.   Results and Discussion

With reference to Section 2 we have generated a set of *measured* sub-pixel displacements for a range of *known* sub-pixel displacements *(nw/10)* where $n = 0, 1, 2, \ldots 10$. This has been calculated for an individual snapshot of the flow field for a single interrogation region of 21 x 21 pixels and a particle image size of 2.8 pixels. We now use the model to simulate a time resolved PIV experiment to generate a time sequence of snapshots where the same range of displacements are measured, but the spatial position of the seeding particle images in the interrogation region is varied randomly between snapshots. For an actual particle image displacement d_a we calculate a measured displacement d_i where $i = 1, 2, \ldots N$ and N is the total number of snapshots in the time sequence of the PIV experiment. We can now define a mean displacement d_m and mean bias error d_b by

$$d_m = \frac{1}{N}\sum_{i=1}^{N} d_i \Leftrightarrow d_b = d_m - d_a \quad (6)$$

Accordingly we can define an rms error σ by

$$\sigma = \left(\frac{1}{N}\sum_{i=1}^{N}(d_i - d_m)^2\right)^{1/2} \quad (7)$$

In this way, we are able to examine the effect of variations in the shape of the correlation peak, which occur, and the corresponding variation in the measured displacement due to the curve fitting routine.

Mean bias error d_b and the rms error σ for the range of sub-pixel displacements and a sequence of 100 snapshots for non-clipped and clipped particle images is shown in Figure 5 (a) and (b). Normalised cross-correlation results are labelled $R_N(m,n)$ and direct cross-correlation results are labelled $R(m,n)$. The trend of the results from Figure 5(a) agree with those of Huang et al. (1997). At 0.5 pixel displacement the rms errors from both $R_N(m,n)$ and $R(m,n)$ are at a maximum for non-clipped particles. It is interesting to note in Figure 5(b), the normalised cross correlation has lower magnitude of errors because this cross correlation routine has less mean bias error due to the normalisation of interrogation region images before correlation. This result agrees with those of Westerweel et al. (1997) and Towers and Towers (2001). Clearly the presence of particle images which are "clipped" by the interrogation region will be a significant variable affecting the correlation peak calculation, because the number of whole particle images that can be contained in an interrogation region is reduced with increasing sub-pixel displacement. This error arises due to the peak finding algorithm. In theory, the correlation function will have a Gaussian profile. When the centre of this profile coincides with a node in the correlation field grid, the digitised correlation function will be symmetric and curve fitting a Gaussian profile to it will produce minimum error. If the digitised profile centre is somewhere in between the nodes, which is most often the case, the curve fitting procedure is less accurate.

It is interesting to note that in practice for a PIV experiment with an interrogation region size of 21 x 21 pixels the laser pulse separation would have been chosen typically to produce a total particle image displacement of 5 pixels. For an actual displacement of 5.5 pixels the results in Figure 5(a) show that for $R(m,n)$ the corresponding rms error (0.0129 pixels) produces an approximate error of 0.24% in the measured displacement over 100 realisations. This means that in a practical flow situation time resolved PIV is capable of contributing a *pseudo*-rms turbulence intensity level of 0.24% and intensity levels below this cannot be measured. However, for clipped particle images, as shown in Figure 5(b), the corresponding rms error of 0.54% was evaluated for $R(m,n)$. It should be highlighted, that the normalised cross correlation has a positive effect when there are clipped particle images, in which the rms level is reduced to 0.50% for an actual displacement of 5.5 pixel.

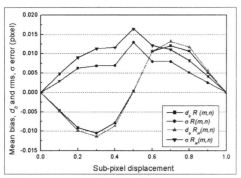

Figure 5(a) Errors for non-clipped particle images   Figure 5(b) Errors for clipped particle images

6.   Conclusions

This paper has described the modelling of a PIV experiment with a view to quantifying the random errors inherent in the technique. The case of a uniform flow across an interrogation region in the absence of any velocity gradients has been examined. The effect of the spatial variation of particle image positions over 100 realisations of the same experiment has been examined over a range of sub-pixel displacements. In addition, the importance of normalising interrogation region images before cross-correlation, as identified by Huang et al. (1997) and Towers and Towers (2001) was examined. Results from Figure 5(b) showed that the normalisation reduces the random error. In the clipped particle images, direct cross-correlation (normally calculated by Fast Fourier Transform) apparent rms turbulence intensities as high as 0.54% can be generated by time-resolved PIV when, in fact, no turbulence is present in the flow.

References

Adrian R J 1997 *Measurement Science and Technology* 8 1393-1398
Goodman J W 1996 *Introduction to Fourier Optics* (New York: Mc-Graw-Hill)
Hart D P 1999 In *PIV '99 Workshop* Santa Barbara, California, USA
Huang H, Dabiri D and Gharib M 1997 *Measurement Science and Technology* 8 1427-1440
Pust O 2000 In *10th Int. Symp. on App. of Laser Techniques to Fluid Mechanics* Lisbon, Portugal
Towers D P and Towers C E 2001 In *4th International Symposium on PIV* Gottingen, Germany
Westerweel J, Dabiri D and Gharib M 1997 *Experiments in Fluids* 23 20-28
Willert C E and Gharib M 1991 *Experiments in Fluids* 10 181-193

Inst. Phys. Conf. Ser. No. 177
Paper presented at 1st Int. Conf. on Optical & Laser Diagnostics, London, 16–20 Dec. 2002
©*2003 IOP Publishing Ltd*

A High Power Single-Mode Tunable Laser for Nonlinear Optical Diagnostics in Combustion

Karen Bultitude and Paul Ewart

Oxford Institute for Laser Science, Physics Department, Oxford University, Oxford, UK

Abstract. Nonlinear optical techniques for combustion diagnostics place stringent demands on the lasers employed. High resolution techniques require narrow bandwidth, preferably single longitudinal mode (SLM), combined with wide tunability and high power. A particular aim of our work is to provide a suitable device for nonlinear optical spectroscopy in the mid-infra-red for diagnostics of hydrocarbon radicals. We report the development of a laser device with the desired properties and its application to high resolution spectroscopy of combustion species.

The system is based on a modeless laser used as a narrow bandwidth amplifier (NBA) of radiation from a widely tunable, SLM, diode laser. When pumped by 200 mJ of energy from a frequency doubled, SLM, Nd:Yag laser the system provides output pulses of up to 30 mJ energy in a transform limited bandwidth of <200 MHz, or 0.006 cm^{-1}. The single mode frequency is continuously tunable over the diode range of 10 nm.

Application of the system to combustion diagnostics is demonstrated by high resolution DFWM spectroscopy of OH in a methane/air flame. Resolution of the spectral lineshape is shown to distinguish effects of Doppler and homogeneous broadening. Applications of the system to other spectroscopic techniques and spectral regions will be reported.

Inst. Phys. Conf. Ser. No. 177
Paper presented at 1st Int. Conf. on Optical & Laser Diagnostics, London, 16–20 Dec. 2002
©2003 IOP Publishing Ltd

Optical Diagnostics for Internal Combustion Engines

H Zhao

Department of Mech. Eng., Brunel University, West London, U.K.

Abstract. The purpose of this presentation is to provide an overview on the instrumentation and techniques for in-cylinder flow and combustion measurements in an internal combustion (IC) engine, in particular the optical diagnostic techniques. It starts with a brief discussion on the aims of in-cylinder flow and combustion measurements and the common thread that ties the different diagnostic techniques. This is followed by a brief discussion on the classification of the experimental techniques and their roles in the engine development. The main part of the presentation then goes on discussing various optical techniques for in-cylinder flow and combustion measurements. The diagnostic techniques will be tied to the characterisation of thermodynamic, chemical and fluid dynamic states of the in-cylinder mixture before, during, and after combustion. Accordingly, the optical diagnostics will be grouped and reviewed as follows: in-cylinder flow measurements, fuel distribution in the combustion chamber, fuel atomisation and vaporisation, combustion and flame visualisation, combustion temperature measurements, detection of combustion products.

This approach ensures that the development of diagnostic techniques is linked to an identified need, as it is the end that justifies the means but not vices versa. This approach is particularly useful for researchers or engineers who are interested in the diagnostic tools that are capable of solving a specific research problem, irrespective of the nature of each technique, whether it is classical or modern. At the same time, by presenting different techniques for the same diagnostic purpose, the limitations of each technique can be understood better for intelligent application and successful outcome.

Inst. Phys. Conf. Ser. No. 177
Paper presented at 1st Int. Conf. on Optical & Laser Diagnostics, London, 16–20 Dec. 2002
©2003 IOP Publishing Ltd

Temporal Measurement of Mass Concentration of Soot Aggregates in the Diesel Exhaust by Two-Color Extinction Method

Takeyuki Kamimoto, Toshiaki Nakajima and Yasushi Kawashima

Tokai University, 1117 Kitakaname, Hiratsuka-shi, Kanagawa 259-1292 Japan

Abstract. A two-color extinction method has been proposed that can measure the temporal mass concentration of soot aggregates in the diesel exhaust. Two laser beams in co-axial alignment transmit a soot loaded exhaust gas flow, and the transmittance at each wavelength is detected simultaneously. The scattering-to-absorption ratio in the extinction coefficient for the soot aggregates was theoretically determined by the transmittance values measured at two wavelengths. The refractive index of the soot aggregates was also determined by a comparison of mass concentrations measured under steady operating conditions by both the extinction and a paper filter methods. Determining the scattering-to-absorption ratio in the extinction coefficient and the refractive index allowed significant improvement in the accuracy of the extinction method.

1. Introduction

Diesel particulate is regarded with suspicion as a pollutant having bad effect on respiratory organs of humankind. Particularly particles having diameters below 100 nm is reportedly accumulated in lung, and hence EPA of USA started a regulation in 1997 to PM 2.5 emission. Diesel particulate is composed of hydrocarbons as sources for secondary particles and soot particles that occupy the major portion of diesel particulate. Current interest is mainly directed toward the measurement of number density distribution of such particles, but measuring the tail-out soot mass concentration is still important to achieve higher efficiency in combustion and improved control of soot emission. Among various measurement techniques, the classical extinction method is apparently a candidate for this purpose, and has been studied by several researchers [1,2]. However, both the effect of scattering on light extinction and the refractive index of soot still remain unclear. The authors propose a novel two-color extinction method to remove these two uncertainties by incorporating the two transmittances measured at two wavelengths and TEM photographs into the data analysis based on the soot aggregates theory [3,4]. This paper will outline the procedures to determine the scattering-to-absorption ratio in the extinction coefficient and the refractive index of diesel soot.

2. Extinction of light transmitting a cloud of soot aggregates

When a collimated light beam radiates a cloud of aggregates suspended in the exhaust gas, the extinction of transmitted light can write as follows by Bouguer-Lambert law.

$$\tau = \exp(-\sigma_{ext} c_m L) \tag{1}$$

where τ is the transmittance, c_m is the mass concentration of soot aggregates (kg/m^3). The specific extinction coefficient, σ_{ext}, (m^2/kg) is composed of specific absorption coefficient σ_{abs} and specific scattering coefficient σ_{sca}.

$$\sigma_{ext} = \sigma_{abs} + \sigma_{sca} \tag{2}$$

$$\sigma_{abs} = \frac{6\pi}{\lambda \rho_p} E(m) \tag{3}$$

Where

$$E(m) = Im\left(\frac{m^2 - 1}{m^2 + 2}\right)$$

$$= \frac{6n\kappa}{\left(n^2 + \kappa^2\right)^2 + 4\left\{1 + \left(n^2 - \kappa^2\right)\right\}}$$

Where $m = n + ik$ is the refractive index of soot, and $\rho_p = 1.86g/cm^3$ [3]. σ_{sca} can be given by the aggregate theory as follows [3,4].

$$\sigma_{sca} = \frac{4\pi F(m)}{\lambda \rho_p} f_n \alpha_p^3 \overline{n_p} \left[1 + \frac{16}{3D_f} \alpha_p^2 \left(\frac{\overline{n_p}}{k_f}\right)^{2/D_f}\right]^{-D_f/2} \tag{4}$$

Where

$$F(m) = \left|\frac{m^2 - 1}{m^2 + 2}\right|^2 \qquad \overline{n_p^1} = \frac{\sum n_{pi} dN_{ai}}{\sum dN_{ai}} \qquad \overline{n_p^2} = \frac{\sum n_{pi}^2 dN_{ai}}{\sum dN_{ai}}$$

$$f_n = \frac{\overline{n_p^2}}{\left(\overline{n_p^1}\right)^2} \qquad \alpha_p = \frac{\pi d_p}{\lambda}$$

f_n is the moment ratio specific to the pdf of aggregate size distribution. d_p is the mean diameter of the primary particles composing aggregates, and $\overline{n_p^1}$ is the average number of primary particles composing the aggregates. k_f is a pre-factor representing the aggregates geometry and D_f is the fractal dimension of the aggregates.

σ_{ext} can write from Eq.(3) and Eq.(4) as

$$\sigma_{ext} = \sigma_{abs} + \sigma_{sca}$$

$$= \sigma_{abs}\left(1 + \overline{\rho_{sa}}\right) \tag{5}$$

Where

$$\overline{\rho_{sa}} = \frac{2}{3}\frac{F(m)}{E(m)} f_n k_f \alpha_p^3 \left(\frac{\overline{n_p}}{k_f}\right)\left[1 + \frac{16}{3D_f}\alpha_p^2\left(\frac{\overline{n_p}}{k_f}\right)^{2/D_f}\right]^{-D_f/2} \tag{6}$$

When $\overline{n_p}$ increases to several hundreds, ρ_{sa} approaches asymptotically to $\overline{\rho_{sa\infty}}$ given by the next equation.

$$\overline{\rho_{sa\infty}} = \frac{2}{3}\frac{F(m)}{E(m)} f_n k_f \alpha_p^3 \left(\frac{3D_f}{16\alpha_p^2}\right)^{D_f/2} \tag{7}$$

3. Experimental and data analysis

A three-liter turbo-charged DI diesel engine (Isuzu-4JX) was used for the experiment. An exhaust gas stream at a flow rate of 50 to 100 liter/min (around 1 % of the total exhaust) was introduced to a cupper tube of 2.17 m long. Two diode laser beams with respective wavelengths of 635 nm and 780 nm were aligned co-axially and pass the exhaust gas in the axial direction through two quartz windows attached in opposite to both sides of the tube. The temperature of the exhaust gas and the surface temperature of the tube were both thermo-controlled at 150 ℃. To avoid soot contamination on the window surface, the temperature of the windows were heated to 250℃. The intensities of transmitted light were detected respectively by two silicon photo-diodes. The digitized data of intensity was analyzed to provide the scattering-to-absorption ratio in the extinction coefficient and the mass concentration of soot aggregates in the diesel exhaust. A small portion of the exhaust gas was injected through an orifice with a diameter of 1.0 mm against cupper meshes for TEM photographs placed 100 mm downstream of the orifice. An image analysis of the TEM photographs provided the characteristics of the soot aggregates.

The TEM photographs showed that typical soot aggregates sampled at a medium load condition at an engine speed of 1500 rpm are composed of hundreds of primary particles of an average diameter of 32.9 nm. The average number of primary particles, $\overline{n_p}$, ranged from 130 to 220 with an average number of 200. f_n was 1.40 for all the data acquired.

To calculate c_m from measured τ, it is necessary to determine $\overline{\rho}_{sa}$ expressed by Eq.(6) in which unknown parameters such as $F(m)/E(m), k_f, f_n, \alpha_p$ and D_f are included. The ratio of $\overline{\rho}_{sa}$ to $\overline{\rho}_{sa\infty}, R$, can be calculated against n_p without any knowledge of $F(m)/E(m)$ and f_n by the next equation.

$$R = \frac{\overline{\rho}_{sa}}{\overline{\rho}_{sa\infty}} = \frac{\overline{n}_p}{k_f} \cdot \left[\frac{\frac{16}{3D_f} \alpha_p^2}{1 + \frac{16}{3D_f} \alpha_p^2 \left(\frac{\overline{n}_p}{k_f} \right)^{2/D_f}} \right]^{D_f/2} \tag{8}$$

Where a widely accepted value of 1.75 was used as D_f [4,5]. $\overline{n}_p = 200$ was assumed from the TEM data. R_1 and R_2 were calculated for $\lambda_1 = 635nm$ and $\lambda_2 = 780nm$, respectively assuming $k_f = 6$ and 9.

To determine $\overline{\rho}_{sa1}$ and $\overline{\rho}_{sa2}$, we need to calculate $\overline{\rho}_{sa\infty1}$ and $\overline{\rho}_{sa\infty2}$ each of which includes parameters, $F(m)/E(m), k_f$ and f_n as shown in Eq.(7). However, it is uncertain to determine the value of each of them, and we put these parameters in a constant, C, as shown below.

$$C = \frac{2}{3} \frac{F(m)}{E(m)} \cdot f_n \cdot k_f \tag{9}$$

Now constant, C, is correlated to the ratio, $\ln \tau_2 / \ln \tau_1$, using Eqs.(3),(7) and (8) as follows.

$$\frac{\ln \tau_2}{\ln \tau_1} = \frac{\sigma_{abs2}}{\sigma_{abs1}} \cdot \frac{1 + \overline{\rho}_{sa2}}{1 + \overline{\rho}_{sa1}}$$
$$= \frac{\lambda_1}{\lambda_2} \cdot \frac{1 + R_2 C \alpha_{p2}^{3-D_f} (3D_f/16)^{D_f/2}}{1 + R_1 C \alpha_{p1}^{3-D_f} (3D_f/16)^{D_f/2}} \tag{10}$$

In calculation, $\alpha_{p1} = \pi \overline{d}_p / \lambda_1$ and $\alpha_{p2} = \pi \overline{d}_p / \lambda_2$ can be given using $\overline{d}_p = 32$ nm and two wavelengths respectively. The correlation between C and $\ell n \tau_2 / \ell n \tau_1$ calculated by Eq.(10) is presented in Fig.1. It is seen in the figure that C can be determined simply from measured τ_1 and τ_2, and is almost independent of k_f assumed. Once C is obtained, $\overline{\rho}_{sa1}$ and $\overline{\rho}_{sa2}$ can be calculated against n_p respectively using R_1 and R_2 determined beforehand. Fig.2 shows the variations of $1 + \overline{\rho}_{sa1}$ and $1 + \overline{\rho}_{sa2}$ with $\overline{n}_p$ with k_f as a parameter at a condition indicated in the legend. It is worth to note that $\overline{\rho}_{sa1}$ and $\overline{\rho}_{sa2}$ can be finally determined as 19 % at 635 nm and 14% at 780 nm respectively for $\overline{n}_p = 200$ and $C = 6$.

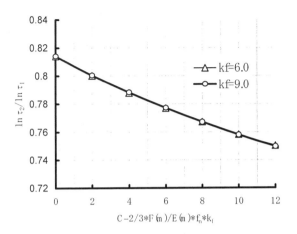

Fig.1 τ_2 / τ_1 versus $C = 2/3 * F(m)/E(m) * k_f * f_n$ with $C = 6$

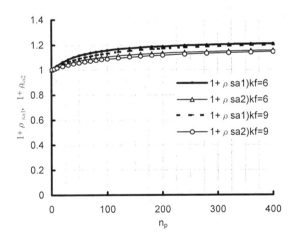

Fig.2 $1 + \rho_{sa1}, 1 + \rho_{sa2}$ versus $\overline{n_p}$ with $C = 6$

4. Result and Discussion

The engine was operated at 1500 rpm, and the temporal variations of transmittances at 635 nm and 780 nm and the engine torque were recorded for 500 second. Before and after measurement, each of the two signal levels corresponding to 100 % transmittance was recorded for calibration. The soot mass concentration at normal condition, c_{mN}, was measured by a smoke meter (AVL: 415S) at steady operating conditions.

Table1 Result of analysis

τ_1	τ_2	$\ell n\tau_2/\ell n\tau_1$	c_{mN}	c_{mEx}	$\sigma_{abs}/10^3$	$E(m)$
0.861	0.891	0.771	14.01	9.54	5.96	0.362
0.869	0.893	0.806	14.01	9.54	5.60	0.339
0.867	0.895	0.777	13.45	9.16	5.93	0.360
0.874	0.900	0.782	13.45	9.16	5.59	0.339
0.840	0.873	0.779	18.75	12.77	5.19	0.315
0.834	0.872	0.755	18.75	12.77	5.41	0.328
0.908	0.931	0.741	9.09	6.19	5.93	0.359

The result of analysis is summarized in Table1. The average value of $\ell n\tau_2/\ell n\tau_1$ is 0.78, and this yields $C = 6$ from the curve shown in Fig.1. c_{mEx} in the table stands for the soot mass concentration at the exhaust gas temperature of 150 ℃ converted from c_{mN} listed also in the table. The values of $E(m)$ calculated from σ_{abs} for 635 nm relate directly to the refractive index, m. The average $E(m)$ of 0.354 at 635 nm is very close to 0.358 that is calculated using the refractive index of $m = 1.55 - i0.78$ proposed by Dobbins et al. [3]. The value of $F(m)/E(m)$ calculated using this refractive index is 0.893, and this gives 6 to C if $k_f = 6.7$. Since k_f ranges from 5.8 to 9, $k_f = 6.7$ is a possible value in this context.

5. Conclusion

1. The scattering-to-absorption ratio in the extinction coefficient was successfully determined by incorporating two transmittance values measured at two wavelengths and TEM data into the soot aggregates theory. The scattering-to-absorption ratios obtained for diesel soot aggregates were 19 % at 635 nm and 14 % at 780 nm.
2. The refractive index of diesel soot was determined from the scattering-to-absorption ratio obtained and the soot mass concentration measured at steady operating conditions. A refractive index of $m = 1.55 - i0.78$ was found most recommendable for the extinction measurement of soot aggregates in the diesel exhaust.

References

[1]   Scherrer H C, Kittelson D B and Dolan D F 1981 *SAE paper* No.810818
[2]   Ressler D M 1982 *Appl. Opt.* **21** 4077-4086
[3]   Dobbbins R A, Mulholland G W and Bryner N P 1994 *Atm. Envir.* **28** 889-897
[4]   Koylu U O and Faeth G M 1992 *Comb.and Flame* **89**140-156
[5]   Puri R, Richardson T F, Santro R J and Dobbins R A 1993 *Comb. and Flame* **92** 320-333

Inst. Phys. Conf. Ser. No. 177
Paper presented at 1st Int. Conf. on Optical & Laser Diagnostics, London, 16–20 Dec. 2002
©2003 IOP Publishing Ltd

Characterization of Mixture Distribution and Combustion in a DISI Optical Engine Under Late Injection Mode

Dong-Seok Choi, Yongseok Lee, Jaejoon Choi and Choongsik Bae

Korea Advanced Institute of Science and Technology
373-1, Kusong-Dong, Yusong-Gu, Daejon 305-701, Korea

Abstract. This study investigated combustion characteristics paying particular attention to a mass flow rate pattern under stratification charge mode. The mass flow rate was changed using two injector drivers with different operating characteristics for an injector. These injection systems were employed in an optical single-cylinder engine. Three different injection timings under late injection modes were set to simulate the stratification charge combustion. To obtain the mass flow rate, the experiments of fuel metering and injection rate measurement were performed with a fixed injection quantity. In-cylinder pressure was measured to analyze combustion characteristics. Initial flame size and its developments were also visualized. In addition, the mixture distribution prior to ignition was obtained using PLIF technique. As the mass flow rate has a higher value, more stable and faster combustion was achieved. The mixture distribution showed more uniform and higher concentration in the bowl near the spark plug as the mass flow rate increases.

1. Introduction

The performance of Direct-Injection Spark Ignition (DISI) engines, especially at part loads, depends on the dynamic characteristics of fuel injection system. In other words, a mass flow rate of the injection system has a significant effect on stratified combustion under light load operating conditions [1, 2]. Hence, the characteristics of the mass flow rate obtained from the fuel injection rate are extremely important in designing the DISI engines.

For stable stratified combustion, the spray characteristics determined by the mass flow rate must provide the proper development of a mixture distribution which will maintain combustion after ignition. In addition, it is necessary to concentrate the appropriate air/fuel mixture around the spark plug for stable flame propagation [3, 4]. It has been considered difficult to achieve stable combustion over a wide range of operating conditions since the permissible range of mixture concentrations is narrow.

The control of the mass flow rate is, therefore, very important since it governs the combustion rate and pollutant formation. Short injection durations less than 1ms are recently required to solve emission problems in two-stage injection strategy [5, 6]. It could be possible to apply the split injection of more than two times as another injection strategy for lean combustion. Thus, the effect of the mass flow rate on the mixture distribution and combustion characteristics needs to be clarified to achieve the stable combustion and reduction of pollutant emissions.

The objective of this study is to investigate combustion characteristics paying particular attention to the mass flow rate patterns under stratification charge mode. For this aim, the mass flow rate was varied using two injector drivers with the different operating characteristics for an injector. These injection systems were employed in an optical single-cylinder engine. Three different injection timings under late injection conditions were tested to simulate the stratification charge combustion. To obtain the mass flow rate, the experiments of fuel metering and injection rate measurement were performed at a fixed air-fuel ratio. In-cylinder pressure was measured to analyze combustion characteristics such as IMEP, COV of IMEP and mass fraction burnt. Initial flame size and its developments were visualized. In addition, the mixture distribution prior to ignition was also obtained using Planar Laser Induced Fluorescence (PLIF) technique. From these experimental results, the effect of the mass flow rate on combustion and mixture distribution was examined under late injection modes.

2. Experimental setup

2.1 Optical engine

The experiments were performed in an optical single cylinder engine modified from a production engine. As shown in Fig. 1, the research engine consists of cylinder head, visualization modules in the extended piston and liner, two connecting plates and cylinder block. A specially designed intake port induces a strong reverse tumble which guides the injected fuel with the help of the special piston bowl to the spark plug. Figure 2 shows the details of a piston top. Flame and fuel inside of the piston bowl were visualized through the bottom view window and optical prism. In the PLIF experiments, this prism plays a role in turning laser sheet horizontally into the inside of the piston bowl. In Fig. 2(a), the observation area was confined by the bottom view window. A high-pressure swirl injector with a cone angle of 60° was used for gasoline direct injection. Table 1 summarizes the engine specification.

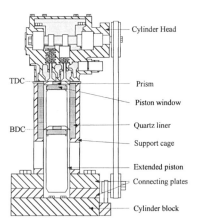

Figure 1 Schematic of an optical engine

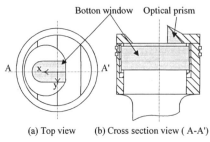

(a) Top view   (b) Cross section view (A-A')

Figure 2 Configuration of piston top

Table 1 Specifications of the optical engine

Engine Type	4 stroke, 4 valve
Bore × Stroke (mm)	85 × 88
Displacement vol. (cc)	559
Combustion chamber	Pent-roof
Compression ratio	9.3
Valve timing	IVO: 6° BTDC, IVC: 46° ABDC EVO:130° ATDC, EVC: 10° ATDC
Injector type	Hollow cone, 60° swirl
Ignition method	TCI

2.2 Measurements

The engine was operated at 800rpm. The coolant temperature was set at 70°C. Iso-octane was used as fuel. Air-fuel ratio was fixed at 35 for all experiments. Injection pressure was set at 5MPa by compressed nitrogen. Three different injection timings (50° BTDC, 60° BTDC and 70° BTDC) were set to investigate the effect of injection timing on fuel stratification. MBT spark timings for each injection timing were found. Combustion analysis was conducted by calculating from in-cylinder pressure through the engine test without the optical modules.

The mass flow rate characteristics of the injector were obtained by independently measuring the relative momentum of the fuel spray and the total mass of fuel injected. The relative spray momentum was measured with a linear force transducer (a Kistler 6123 piezoelectric pressure transducer) placed at 1.5mm in front of the injector tip.

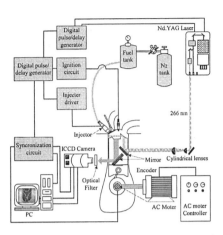

Figure 3 Experimental setup for the visualization of flames and mixture distribution

Experimental setup for the flame visualization and PLIF measurements is shown in Fig. 3. The PLIF images were digitally recorded with an intensified-CCD (ICCD) camera that provided 640 by 480 pixel images at a resolution of 8bits. For the flame visualization, the bottom view window and 45° degree mirror were used. For the PLIF experiments, the laser light sheet was induced inside of the piston bowl by the 45° degree mirror and optical prism. The fourth harmonic of the Nd:YAG at 266nm was used to excite dopants. Base fuel and dopant used were iso-octane and 3-pentanone, respectively. Their mixing ratio was 80% iso-octane and 20% 3-pentanone by volume.

3. Results and discussion

3.1 Mass flow rate

Mass flow rate measurements were performed to determine the injection duration required to meter a given amount of fuel for an injector with two different injection drivers. The results of the experiments, on a per injection basis, are shown in Fig. 4. Each curve represents a different injector driver. To determine the injection duration required for the preset air-fuel ratio, the metering data in Fig. 4 were curve fit. The amounts of injection for types A and B were measured by weighing over 1000 injections at short injection times (less than 1.0ms) and 100 injections for the longer injection time. The equations noted on Fig. 4 represent linear regressions of the data. Correlation coefficients were greater than 0.99 in all cases.

In addition to metering experiments, injection rate measurements were performed using the method introduced by Naber et al. [7]. Rate of injection experiments give a time history of fuel mass emitted during an injection event. Figure 5 shows the results of the injection rate measurements. The vertical dotted lines on each curve represent the end of injection control signal. The time lag between the control signal (time zero) and the initial climb in the rate of injection is due to a delay associated with each injector driver. The time lags for two injector drivers (types A

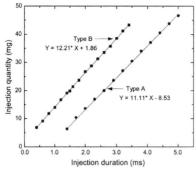 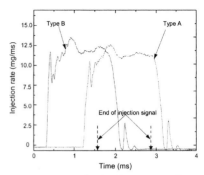

Figure 4 Fuel metering as a function of injection duration   **Figure 5** Rate of injection curves for two different injector drivers

and B) were found to be approximately 0.26ms and 1.25ms, respectively. Each rate of injection curve exhibits transient behavior at the beginning and end of the injection event. Between these two transient features the rate of injection is nearly constant. The first peak could result from a leading mass followed by the main spray [8]. The slope of the initial climb of type B is steeper than that of type A. This indicates that type B has a bigger mass flow rate than type A at the early stage of injection. In addition, the time approaching the steady mass flow rate of type B is faster than that of type A. Another peak in injection rate curve after the end of injection is likely due to some needle bounce. For the stratified combustion, the initial mass flow rate is very important since the mixing time between the injected fuel and intake air is not sufficient. Thus, the effect of two different mass flow rates on combustion and mixture distribution was investigated in the following sections.

3.2 Combustion characteristics

3.2.1 Analysis of in-cylinder pressure

To evaluate the effect of the difference of the initial mass flow rate, the combustion analysis was carried out to estimate IMEP, Coefficient of Variability of IMEP (COV_{IMEP}) and mass fraction burned (MFB) from in-cylinder pressure data. Figure 6 shows IMEP and COV_{IMEP} as a function of injection timing for the injector drivers of types A and B. Error bars represent COV_{IMEP}. For both types, MBT ignition timing was at 25° BTDC. IMEP of type A decreases with the injection timing advance while that of type B increases. However, IMEP of type B has higher values than that of type A for all injection timings. This result indicates that the higher initial mass flow rate leads the bigger IMEP. To examine combustion duration, MFB was calculated. The results of the analysis of mass fraction burned were summarized in Table 2. Negative and positive values mean before top dead center and after top dead center at a compression stroke, respectively. The slopes of the mass fraction burned for type A is smoother than those for type B with a higher mass flow rate. In particular the 50% mass fraction burned for best cases of both types is well positioned at 5-10° ATDC. As the injection timing is retarded, the burn duration of type A is

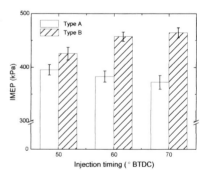

Figure 6 IMEP with COV of IMEP for two different drivers

Table 2 Combustion durations for two different mass flow rates

Injection Timings MFB	Type A			Type B		
	50° BTDC	60° BTDC	70° BTDC	50° BTDC	60° BTDC	70° BTDC
10 % (°CA)	-0.7	-0.8	1.4	-4.1	-4.6	-0.5
50 % (°CA)	9.8	10.0	13.3	3.4	4.1	8.9
90 % (°CA)	22.0	21.7	25.0	15.5	15.9	18.4
Burn dur. (°CA)	22.6	22.6	23.6	19.6	20.5	18.9
Slope	17.0	17.1	16.4	19.7	18.8	20.7

shorter while that of type B becomes longer. Thus, the case of type B with higher mass flow rate showed faster combustion, higher IMEP and lower COV_{IMEP} as shown in Fig. 6.

3.2.2 Initial flame development

To observe flame shape and size at the initial stage of combustion, flame visualization by self luminescence at this stage was performed with ICCD camera. Two representative cases, which are best IMEP cases for both types of injection driving units, are shown in Fig. 7. Five images for each case were averaged and flame front was detected by an image processing. One is the case of 50° BTDC injection timing for type A. The other is the case of 70° BTDC injection timing for type B. As shown in Fig. 6, IMEP of the latter case is higher than that of the former case. Each number indicates crank angle degree before TDC. The flame propagates downwardly. The flame speed of type B is slightly faster than that of type A. At the early stage from 23° BTDC to 19° BTDC, the flame size for type B is bigger than that for type A and the flame shape is more regular. After the early stage, flame spread outside of the observation area. These results imply that the case with faster mass flow rate has more sufficient mixing time resulting in stable combustion.

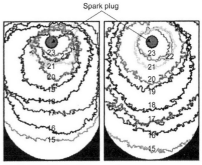

(a) Type A   (b) Type B
50° BTDC injection   70° BTDC injection
Figure 7 Initial flame development

3.3 Fuel distribution

To examine mixture distribution prior to combustion, PLIF technique was employed. The image processing was done on a set of ten images for each condition. The images were scaled for laser power and camera gain, normalized using the measured vertical sheet profile and then averaged. Figures 8 and 9 show the relative fluorescence intensity along the vertical center line and the horizontal line with a spark plug. To compare each mixture distribution, the intensity was normalized by the maximum value of all cases. Due to inducing the laser sheet through the bottom window and prism, the position of laser sheet varies with crank angle. Figure 8 (a) and (b) are same condition as shown in Fig. 7.

For type A, the high intensity de to the fuel injected appears at 46° BTDC. As the crank angle advances, the dense mixture moves to the spark plug due to the piston wall and intake tumble motion. The intensity near spark timing was about 0.3. For type B, no fuel was observed at 46° BTDC owing to the retarded injection timing. Due to the same mechanism, the mixture concentrates near the spark plug. However, the intensity near spark timing was about 0.5, which is bigger around 1.6 times of type A. This results in higher IMEP value as described in 3.2.1.

76

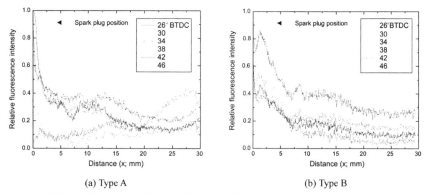

(a) Type A　　　　　　　　　　　　　(b) Type B

Figure 8 Fluorescence intensity along the vertical center line through spark plug

Although the graph of horizontal intensity profiles is not shown in this paper, the intensity for type B distributes more uniform and is higher than that of type A. These results imply that more stable stratified combustion process could be achieved.

4. Summary

The effect of the mass flow rate on stratified combustion was investigated by the analysis of combustion characteristics, flame visualization and PLIF measurements indicating the mixture concentration distribution. The main findings are summarized as follows:

(1) As the mass flow rate has a higher value, more stable and faster combustion was achieved.
(2) The initial flame size and shape for higher mass flow rate are bigger than that of lower mass flow rate.
(3) The mixture distribution showed more uniform and higher concentration in the bowl near the spark plug as the mass flow rate increases.

Acknowledgments

This research was supported by NRL (National Research Laboratory) scheme, Ministry of Science and Technology, Korea.

References

[1] Zhao F, Lai MC and Harrington DL 1999 *Prog. Eng. Comb. Sci.* 25 437-562
[2] Cousin J, Ren WM and Nally S 1998 *SAE Paper* 980499
[3] Tabata M, Kataoka M, Tanaka T and Yamakawa M 2000 *SAE Paper* 2000-01-0240
[4] Shayler PJ, Jones ST, Horn G and Eade D 2001 *SAE Paper* 2001-01-3671
[5] Miyamoto N, Ogawa H, Shudo T and Takeyama F 1994 *SAE Paper* 940675
[6] Harada J, Tomita J, Mizuno H, Mashiki Z and Ito Y 1997 *SAE Paper* 970540
[7] Naber JD and Siebers DL 1996 *SAE Paper* 960034
[8] Parrish SE and Farrell PV 1997 *SAE Paper* 970629

Inst. Phys. Conf. Ser. No. 177
Paper presented at 1st Int. Conf. on Optical & Laser Diagnostics, London, 16–20 Dec. 2002
©*2003 IOP Publishing Ltd*

An Investigation of the Effect of Nozzle Characteristics on Liquid Atomisation in a GDI Injector

Y C Khoo and G K Hargrave

Wolfson School of Mechanical and Manufacturing Engineering, Loughborough University, Loughborough, Leicestershire, LE11 3TU,UK.

Abstract. Experimental data is presented for the internal flow structure of 'real size' GDI injectors. Four optical nozzles were used, with inlet chamfered angles varying from 0° to 60° and a Length/Diameter (L/D) ratio of 4. The internal flow structure was studied with a combination of high-speed flow visualisation and Particle Image Velocimetry (PIV). These techniques provided qualitative and quantitative information of nozzle flow for fuel supply pressures varying from 20-50bar. From the PIV data, a simulation model has been developed using CFD codes to aid the understanding of the internal flow processes to assist the future development of GDI injectors. Preliminary results taken from the computational models presented show good agreement with the experimental data.

1. Introduction

The study of Gasoline Direct Injection (GDI) systems for use in internal combustion engines is well documented. The implementation of more stringent legislation for vehicle emissions by Government bodies has increased the interest in GDI technology as a method of improving efficiency and emissions. A GDI engine has the ability to control the precise amount of fuel injected into the combustion chamber, allowing better control of engine performances and lower fuel consumption.

One of the main components that make GDI such an attractive option in terms of higher efficiencies and power output is the GDI injector. Need for fundamental knowledge of spray formations and characteristics is vital for the development of GDI injection systems. Tokuoka et al (1991) and Zhang et al (1991) both showed the influence of the GDI swirl atomiser on the microscopic behaviour of fuel droplets generated at low pressure. With the advancement of laser diagnostic tools, Wigley et al (1998) was able to quantify the hollow cone spray of a GDI injector. Comer et al (1999) used Computational Fluid Modelling (CFD) tool to predict the external flow from the GDI injector. Similarly, experimental and modelling techniques were employed by Arcoumanis et al (1999) and Gavaises et al (2002) to predict the development of the hollow cone spray and the internal flow structure within the GDI injector. However, limited open literature is published on the characteristics of the inlet geometry and it effects on the spray of the GDI injection. Schmidt (1997) investigated the cavitation effects in sharp edge inlet nozzle and successfully predicted the velocity data. Allen et al (2000) investigated the change in the nozzle radii on the generation of a hollow cone spray. He also showed that the removal of cavitation in the flow field enhanced the ability to collect Particle Image Velocimetry (PIV) data.

The aim of this paper is to present the experimental results on the internal flow structure of the nozzles, and the effects of changes in orifice geometries. This includes the input pressure characteristics, internal flow characteristics and the velocity before exiting into the combustion chamber, which is not well documented. PIV provides qualitative and quantitative data for the development and validation of single-phase CFD code.

2. Experimental

2.1 Optical Rig

An optical rig has been designed to allow optical diagnostic techniques to be employed to investigate internal flow structures in a laboratory based environment. The schematics of the two types of optical nozzles layout were shown in Figure 1. All the nozzles used have a L/D ratio of 4.

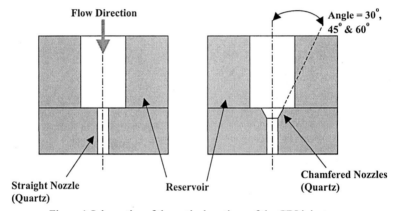

Figure 1 Schematics of the optical sections of the GDI injectors

The working fluid used in the experiment was white spirit, which has a density, viscosity and refractive index similar to that of Gasoline fuel. It is also available in low, regular and high flash versions, the latter having a vapour pressure very similar to gasoline. The fluid was stored in an accumulator and pressurised by nitrogen gas to the required pressure before each controlled injection. A control valve was then attached to the head of the optical rig to allow manual actuation of each injection process.

2.2 Laser Diagnostic Techniques

In this paper, two optical diagnostic techniques were used to analyse the flow field. High Speed Flow Visualisation (HSFV) provided a detailed analysis of the flow characteristics. Fluorescence Particle Image Velocimetry (FPIV) technique was employed to provide quantitative data for characterising the flow structures.

The HSFV technique consisted of a Copper Vapour Laser, 511nm wavelength, synchronised to a Kodak 4540 High Speed Motion Analyser at 9kHz. The laser source provides back illumination through a filter onto the optical rig and allowed the flow to be captured by the imaging camera. These images were then stored as movie file or as individual frames and be analysed.

The FPIV system shown in figure 2 consists of two Nd:YAG Laser systems, with 532nm wavelength, and Kodak ES 1.0 CCD camera with 1008 pixels x 1018 pixels resolution. The twin Nd:YAG lasers were aligned to illuminate the 1μm size fluorescence particles (620nm emission) that were introduced into the flow to act as flow tracers. When the system was triggered by the actuation of a control valve, the CCD camera recorded the light scattering by the fluorescent particles with a pulse separation of 1μsec between frames. A filter is placed in front of the camera to prevent any speckles from affecting the images recorded by the CCD camera.

The FPIV images were then analysed using cross-correlation method with a 64 x 64 pixels interrogation region and a 50% overlap. The analysis region is a 6mm x 6mm square section using Gaussian Fit Algorithm. These FPIV results were be used to validate CFD modelling data.

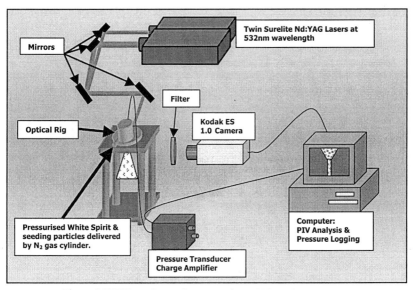

Figure 2 PIV Experimental set-up using 2 Nd:YAG Lasers.

2.3 Mathematical Model Description

A single-phase 3D CFD modelling technique was constructed to provide an analytical description of the transient behaviour inside the nozzle as a function of the different inlet orifices. CFD codes for the simulations were produced using FLUENT Inc. software. A 3D numerical grid was generated to look at the full size nozzle flow characteristics. This provided better quantification of the flow in the nozzle as compared to 1D or 2D geometry. The operating pressure and the fluid simulated were exactly the same as experimental for each controlled injection. Although 3D analysis was used, single plane numerical results was presented in this paper.

3. Results and Analysis

HSFV provides a detailed study on development of the flow field in the nozzles. Figure 3 shows a sequence of straight inlet nozzle's images extracted from the high-speed camera.

The injection was actuated at 30bar pressure with white spirit as the fluid. Cavitation started to develop as the pressure reaches peak pressure. Its formation was due to the presence of low pressure region that occurs near the sharp inlet nozzle. These low pressure region could fall below the vapour pressure of the fluid, thus creating a *vena contracta* effect. Schmidt (1997) found that his flow model had an effective area that was significantly smaller than the nominal nozzle area due to the *vena contracta* effect generated by cavitation. However, cavitation formation was not only sensitive to the change in geometry of the nozzle but also the surface imperfections in the nozzle shape as highlighted by Allen et al (2000). He showed that any surface imperfections created localise cavitation in his GDI nozzles.

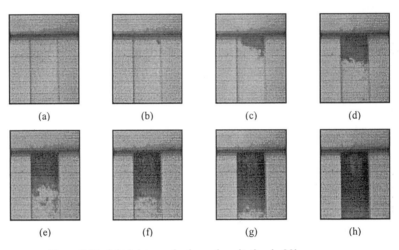

(a)	(b)	(c)	(d)
(e)	(f)	(g)	(h)

Figure 3 Straight inlet nozzle showed cavitation in 30bar pressure.

Although cavitation induced hydrodynamic pressure drop contributes significantly to the liquid atomisation, this renders PIV technique to quantify the flow in the straight inlet nozzle useless. However, Allen et al (2000) also showed that once surface imperfections were removed from his GDI nozzles, good quality velocity validation could be achieved. This led to the development of the chamfered inlet nozzles to allow us to quantify the velocity.

3.1 Velocity Flow Field

Chamfered inlet nozzles were used to look at the effects of the change in inlet orifice and it effects on velocity flow in the nozzle. Figure 4 represents the experimental and numerical simulation velocity profiles of a 45° chamfered inlet nozzle at 30bar pressure. The images showed the flow from the chamfered inlet to the 1mm nozzle and the colour coded bar show the various velocities vectors that was generated in the flow.

From the experimental data showed, the velocity field in the chamfered inlet was much lower before entering the 1mm nozzle. The velocity difference between the chamfered inlet and the 1mm nozzle was about 20 m/s. This similarity was also observed in the numerical simulation results presented. It was also noted that the velocity field profiles were evenly distributed as compared to the experimental. Low velocity field profiles could be seen near the walls on both images, more so with the numerical simulation.

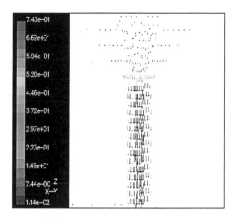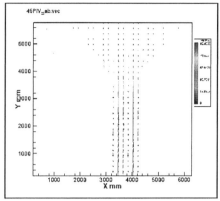

Figure 4 Numerical and experimental velocity vectors with 30bar pressure.

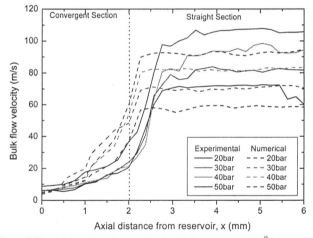

Figure 5 Experimental and numerical velocity profile of 45° inlet nozzle.

Figure 5 shows a comparison of the experimental and numerical data for the axial variation in bulk flow velocity for the 45° inlet chamfered nozzles. These velocity flow profiles show a high degree of similarity between experimental and numerical. At the 2mm point of the nozzle length showed a surge in velocity flow. This was due to the change in flow from the chamfered inlet nozzle to the 1mm diameter nozzle. Once the flow has entered the straight section of the nozzle, the bulk flow velocity remains almost constant. From the bulk flow velocity plots, the numerical predictions are seen to overestimate the velocity in the convergent nozzle section, and underestimate by approximately 10% through the straight nozzle sections. However, qualitatively the agreement is very good.

The variation in maximum flow velocity with fuel supply pressure is presented in figure 6. The agreement between the numerical and the experimental for the 30° inlet chamfer is very good, with deviation for the 45° and 60°. This was also observed when comparing with

Schmidt et al (1999) and Allen et al (2000). Although Schmidt (1999) numerical results was applied on an inlet nozzle with a radius of 0.025 and Allen et al (2000) experimental results used a 45° chamfered nozzle with smooth inlet, it velocity profiles were similar to those shown in Figure 6.

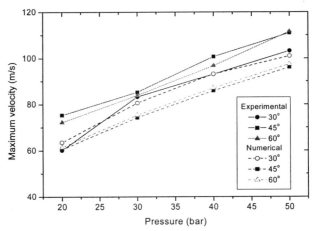

Figure 6 Maximum velocity of experimental and numerical results.

4. Conclusions

HSFV and FPIV were shown to provide experimental data to quantify the flow for a straight inlet and 3 different inlet chamfered nozzles.

HSFV of straight inlet nozzle showed the development of cavitation in the nozzle flow. Undesirable effects of cavitation that occurred in the straight inlet nozzle prevented the use of PIV technique to obtain any useful velocity results. Therefore, future investigation will involve using 2D nozzle to quantify the velocity flow field.

FPIV technique and numerical modelling also produced comparable velocity data for the different chamfered inlet nozzles. The experimental velocity profiles recorded were about 10% higher than numerical results.

References

Allen, J., Hargrave, G.K., 2000, 16th CLASS, Darmstadt, Germany.
Arcoumanis, C., Gavaises, M., Argueyrolles, B., Galzin, F., 1999, SAE 1999-01-0500.
Gavaises, M., Abo-serie, E., Arcoumanis, C., 2002, Direct Injection SI Engine Technology 2002, 2002-01-1136.
Schmidt, D.P., Corradini, M.L., 1997, Atomization and Sprays Vol.7 (No.6).
Schmidt, D.P., Rutland, C.J., Corradini, M.L., 1999, Atomization and Sprays Vol.9 (No.3).
Tokuoka, N., Yamaguchi, Y., Takada, M., Zhang, F., 1991, ICLASS'91, Gaithersburg (USA), Paper 21.
Wigley, G., Hargrave, G.K., Heath, J., 1998, 9th ISALTFM, Lisbon, Portugal, Paper 9.4.
Zhang, F., Tokuoka, N., Yamaguchi, Y., Takada, M., 1991, ICLASS'91, Gaithersburg (USA), Paper 22.

Inst. Phys. Conf. Ser. No. 177
Paper presented at 1st Int. Conf. on Optical & Laser Diagnostics, London, 16–20 Dec. 2002

Development of a Wall Film Thickness Measuring Device

Ronny Lindgren, Richard Block, Ingemar Denbratt

Chalmers University of Technology, Department of Thermo and Fluid Dynamics
SE-412 96 Göteborg, Sweden

Abstract. The choice of technique developed for this study is based on Laser Induced Fluorescence (LIF). The light source consists of a pulsed Nd:YAG laser using the fourth harmonic wavelength of 266nm. The fuel used in the study is iso-Octane with 3-pentanone added as a tracer due to its similar physical properties. Optical fibres are used in order to achieve optical access. One fibre is used for excitation and six fibres are used to collect the emitted light. A photomultiplier tube is used to detect and analyse the emitted light.

The developed method is tested and calibrated to tracer concentration, liquid temperature, PMT gain and change in excitation light energy. A final curve used for obtaining the film thickness is created.

1. Introduction

The use and development of the direct injected stratified charge (DISC) spark ignited (SI) engine has increased in recent years. The purpose of this engine is to combine the advantages from the spark ignited engine with the advantages from the compression ignited engine in order to achieve more efficient engines. There are different ways of obtaining the stratified combustion. The first generation of DISC engines uses the spray and air motion in combination with a bowl in the piston to direct the fuel air mixture towards the spark plug. Investigations have shown that one of the drawbacks with this system is that the spray impinges on the surface with a fuel film build up as result. This fuel film leads to decreased engine efficiency and increased formation of emissions and soot.

To further develop the DISC engines computational fluid dynamics (CFD) plays a central role. Simulations with sprays impinging on surfaces often fail to provide sufficient accuracy. The main reasons for this are both the lack of models describing the phenomena and the lack of experimentally validated data. To make CFD simulations more accurate there is a need of increased knowledge of the physics of the film build up. This in turn requires a method that measures wall film thickness, which is able to operate under realistic engine conditions, to be developed. It is also important that both the temporal and spatial resolution is high. The accuracy of the method has to be in the order of a few micrometers.

2. Experimental set-up

2.1 Fundamentals

Laser Induced Fluorescence is a well-known technique in combustion research [1]. The general principle is to excite a specie by light of a certain wavelength and then measure the fluorescence light that is emitted from the molecules of interest. The wavelength of the emitted fluorescence light is often shifted, called Stokes-shift, and can easily be separated from the excitation wavelength with filters.

When applied to fluid film measurement Lambert-Beers law gives the relationship between the fluorescent intensity and the liquid film thickness. The expression for the intensity of the emitted fluorescence becomes [2]:

$$I_{flour} = QI_o(1 - \exp(-\alpha c \delta)). \tag{1}$$

Here Q is the quantum efficiency, I_0 the exciting intensity, α molar absorption coefficient, c is the concentration of absorption species and δ is the wall film thickness. For small exponents in the series expansion of the exponential function, the terms of higher order can be neglected. Then the expression for the fluorescence light can be written as

$$I_{flour} = QI_o \alpha c \delta. \tag{2}$$

Thus for low concentrations and small thickness the intensity is proportional to the film thickness. The main advantage with this method is the ability to measure the actual film thickness exposed to the laser light. This can be interpreted, through calibration, as a film thickness. The parameters Q, α, c and the response function of the photomultiplier are either unknown or depending on the temperature, which has to be regarded in the calibration procedure.

2.2 Fuel and tracer

The fuel used in this study is iso-Octane (2,2,4-trimethylpentane, C_8H_{18}). This fuel is commonly used in studies for gasoline because its boiling point is 99°C, which is close to the 50% evaporation point of gasoline. One difference between gasoline and iso-Octane is that iso-Octane does not fluorescence naturally when it is exposed to UV light. Therefore a tracer must be added. The choice of tracer is dependent on a number of different demands, such as absorption at used excitation wavelength, satisfactory quantum yield, good solubility in the fuel, sufficient Stokes shift and more. One of the more important demands is that the tracer shows similar evaporation properties as the fuel.

An extensive analysis of different iso-Octane/tracer mixtures has been carried out by Neij et al. [3]. According to [3] 3-pentanone (diethyl ketone, $[CH_3CH_2]_2CO$) is the tracer that best fulfils the demands of this study. The boiling point of 3-pentanone is 102°C which is close to the boiling point of iso-Octane. The fluorescence spectrum for 3-pentanone reaches from 330nm to 600nm with a maximum at 430nm. The absorption spectrum is in the range 220nm to 330nm with a maximum at 280. With excitation laser light in the range 250nm to 320nm the quantum yield shows no change [3], which makes it possible to use any laser with a wavelength in that range.

2.3 Optical set-up

Figure 1 shows a sketch of the measuring device set-up. The light source consists of a Spectra-Physics Quanta-Ray LAB 170/10 pulsed Nd:YAG laser using the fourth harmonic wavelength of 266nm. This light source combined with 3-pentanone as tracer results in a Stokes shift that is sufficiently large to easily separate the excitation laser light from the fluorescence.

The optical access is achieved with fibre optics. One fibre is used for excitation and six fibres are used to collect the emitted light. The fibres are mounted in a small probe with a quarts window which allows optical access through the wall. The probe is fixed to the wall with a screw and is positioned parallel to the wall. This allows the probe to easily be moved to another measurement position.

Figure 1: Image showing the experimental set-up with laser and probe.

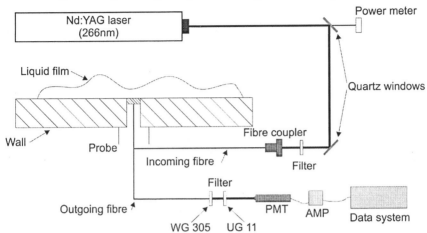

The laser light is focused into the fibre with an OZ Optics HPUC laser-to-fibre coupler. This component is a standard component commercially available. The coupler works by aligning a focusing lens between the laser and the end of the optical fibre. The fibre is mounted to the coupler with a standard SMA contact to make it easier to replace and repair the fibre if it becomes damaged by high laser power. Before the laser light is focused into the fibre the power is decreased by using the reflections from quartz windows. In theory this would decrease the laser power to 4%-8% of the incoming power (4% of the incoming laser power is reflected from each side of the window). Thereafter the light is filtered so only light with a wavelength of 266nm is used.

The optical probe consists of housing around the seven fibres. The excitation light is transported into the measuring volume with one 200µm fibre, mounted in the middle. The fluorescence light is collected with six 200µm fibres mounted coaxially around the incoming fibre. The fibres ends are protected with a small quartz window on the top of the probe. The seven fibres are fixed in the probe with epoxy glue.

The advantage of using a probe, that is easy to move to another measuring position, is that there only is a need of one calibration. Since the probe is not disassembled when moved, the calibration is still valid. The disadvantage is that the probe size becomes a bit larger and makes larger changes to the local surface structure at the measuring position.

The emitted fluorescence light is analysed with a Perkin Elmer MH944 photomultiplier tube (PMT). Before the PMT the emitted light passes through two filters. The first filter is a SCHOTT WG 305, which is used to decrease the influence of the excitation laser light. The second filter is a UG 11 filter that minimises the disturbance from the surrounding light sources such as electric light used to illuminate the laboratory. The signal from the PMT is amplified before it is sampled with a data acquisition system.

3. Calibration

The expression for the film thickness as a function of the intensity of the emitted light is dependent of different system settings and of the surrounding conditions where the measurements are performed. Therefore a calibration must be carried out. According to Neij [3], Fujikawa et al. [4] and Frigo [5] the fluorescence intensity of gas phase 3-pentanone is dependent on the temperature. The changed signal is most probably due to that the absorption spectrum for 3-pentanone is shifted to higher frequencies with increased temperature. This means that the magnitude of the absorption at a certain wavelength is changed giving a change in fluorescence intensity. However, no information on the temperature dependence for the fluorescence signal from a liquid is found. Therefore a temperature variation is included in the calibration procedure.

Information about the pressure dependence on the fluorescence from a liquid can not be found in literature. The investigations that could be found have been carried out on gas phase 3-pentanone and indicates that the intensity of the fluorescence signal is increased with increased pressure [4, 5]. The pressure dependence might be explained with two mechanisms. According to Fujikawa et al. [4] the increase in pressure gives an increasing rate of vibration rotational relaxation from the excited state to lower lying levels with higher fluorescence quantum yield. This relaxation endures to some pressures where the fluorescence quenching by oxygen leads to a decrease in signal intensity. The other mechanism can be that the molecule density of the tracer is increased because of increased gas density, which also leads to higher quantum yield.

When applied to the liquid phase the oxygen quenching can be regarded unimportant since the amount of oxygen in liquid is very small. The change in tracer concentration because of pressure change can be ignored when the liquid is considered as incompressible. Therefore, the pressure effect on the fluorescence intensity is ignored in the calibration process.

A small calibration unit is constructed to do the calibrations outside the spray chamber. A polished aluminium surface with possibility of heating is used. The probe is mounted to the surface and a quartz window is fixed above it. It is possible to adjust the distance between the probe and the quarts window with shims. The probe used during the calibration is the same that is mounted to the wall in the spray chamber and therefore the calibration is valid for all cases. Arrangements are done so the fuel tracer mixture can flow through the measuring gap and the calibration is carried out with fresh mixture.

4. Results

The objective with the calibration is to find the sensitivity between a certain film thickness and fluorescence signal. The sensitivity of the fluorescence signal was tested to tracer concentration, temperature, excitation energy, and PMT gain setting. In all figures the fluorescence intensity is described with the voltage from the PMT.

Two different concentrations of tracer were tested, 3% and 6%. The intensity from a given film thickness is increasing with increasing tracer concentration, see figure 2a. The increase in signal was lower than the increase in tracer concentration. Therefore, to keep the tracer influence to a minimum, a concentration of 3% (v/v) 3-pentanone is used for measurements.

Five different temperatures on the mixture were tested. These were 288K, 309K, 325K, 348K and 363K. It is clearly shown that the fluorescence intensity is decreasing with increasing temperature, see figure 2b. It is also shown that the decrease in signal is independent of the fuel film thickness, i.e. the slope of the curves for different film thickness is almost the same. During measurements it is impossible to measure the exact temperature of the liquid film. Therefore, the temperature dependence must be regarded as an uncertain factor.

The sensitivity of the PMT can be tuned. Changing the gain is done with the high voltage of the PMT. If the sensitivity on the PMT is too high the noise is dominating the signal. If the gain is too low the signal output is too low to record. In figure 2c different gains have been tested. When the curves are investigated closer it is observed that the shape for different gains is slightly changed. That means that a calibration must be performed for the gain used for measurements. The optimum setting were where the noise level was low compared to the fluorescence signal.

The fluorescence signal is sensitive to the energy of the excitation pulse. As can be seen in figure 2d an increase in excitation energy increases the fluorescence intensity. It is therefore very important to keep the excitation energy constant. However, higher energy levels were not tested due to the risk of exceeding the damage threshold of the fibre.

Figure 2: The results from the sensitivity analysis. Figure a) tracer concentration sensitivity, fig. b) temperature sensitivity, fig. c) PMT gain sensitivity and fig. c) sensitivity to excitation energy.

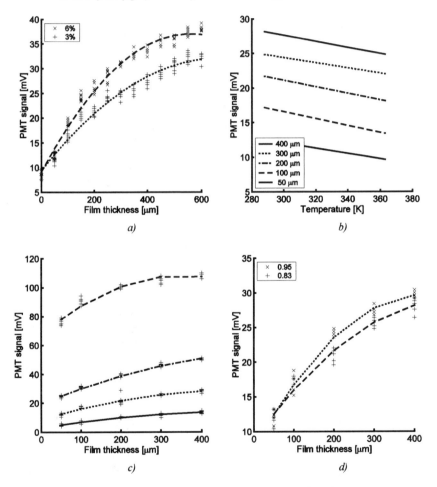

From all the different conditions tested the best combination for measurements were found to be, excitation laser power 0.95mJ/pulse and the PMT high voltage 950V. The calibration curve was obtained with a liquid temperature of 288K. The film thickness is varied from 0μm to 600μm with an increment of 50μm. The final curve is shown in figure 3.

88

As can be seen the curve is close to a second grade expression. This can probably be explained with the fact that the laser illuminated area increases with the distance raised to the power of two, the excitation laser intensity is decreasing at the same rate.

Figure 3: Final curve to be used as reference for at liquid film measurements.

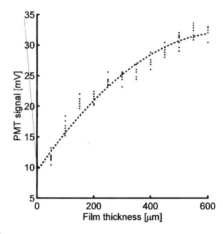

5. Acknowledgements

The authors would like to thank Dr. Michael Försth at Department of Experimental Physics, Dr. Johan Engström at Volvo Technical Development and Dr. Joachim Gronki at University of Heidelberg for the support during the development of this device. For help with producing hardware, Lars Jernqvist, Uno Hansson and Morgan Svensson are gratefully acknowledged.

References

[1]   Eckbreth, A.C. (1996) *Laser Diagnostics for Combustion and Species* (3rd ed.). Amsterdam: Gordon and Breach Science Publishers SA

[2]   Senda, J., Ohnishi, M., Takahashi, T., Fujimoto, H., Utsunomiya, A., Wakatabe, M. (1999) *Measurement and Modelling on Wall Wetted Fuel Film Profile and Mixture Preparation in Intake Port of SI Engine.* SAE Technical Paper 1999-01-0798

[3]   Neij H. (1998) *Development of Laser-Induced Fluorescence for Precombustion Diagnostics in Spark-Ignition Engines.* Lund Reports on Combustion Physics LRCP-41, Lund Institute of Technology

[4]   Fujikawa, T., Hattori, Y., Akihama, K. (1997) *Quantitative 2-D Fuel Distribution Measurements in an SI Engine Using Laser-Induced Fluorescence with Suitable Combination of Fluorescence Tracer and Excitation Wavelength.* SAE Technical Paper 972944

[5]   Frigo, S. (1993) *Evaluation of Fluorescing Tracers for Hexane and Isooctane.* Report # 1981, Department of mechanical and aerospace engineering, Princeton University

Inst. Phys. Conf. Ser. No. 177
Paper presented at 1st Int. Conf. on Optical & Laser Diagnostics, London, 16–20 Dec. 2002
©2003 IOP Publishing Ltd

Simulating the effect of multiple scattering on images of dense sprays

M C Jermy , A Allen and A K Vuorenkoski

Department of Optical and Automotive Engineering, School of Engineering, Cranfield University, Cranfield, Beds, MK43 0AL, UK

Abstract. When using optical diagnostics to measure sprays and other two-phase flows it is common to assume that each detected photon has scattered from only one droplet or particle. However in dense particle fields the number which have scattered more than once may be significant. These multiply scattered photons carry information about more than one region in the spray and as a result the measurement or image becomes difficult to interpret and the accuracy of the measurement is reduced. We have written a flexible Monte Carlo photon transport simulation code capable of handling any 3D geometry and particularly suited to two-phase flows. The code has been used to simulate photon transport in typical dense spray imaging experiments, to quantify the error introduced by multiple scattering. The results show that up to 50% of the photons reaching the camera are multiply scattered. Under the conditions modelled, camera aperture is ineffective in reducing the contribution of multiple scattering to the image. The particles modelled here are non-absorbing, and their scattering is forward dominated. As a result multiple scattering causes little blur or position-dependent attenuation in the images. For particles nearer the Rayleigh limit, forward scattering is not so dominant and blur may be more serious. For thicker systems and systems with absorbing particles the effects are also expected to be more serious.

1. Introduction

Light scattering techniques using visible and near visible wavelengths are popular for studying particulate systems such as sprays, clouds and powders. Scattering techniques can be non-intrusive i.e. without physical probes that disturb the sometimes delicate hydra- and aero-dynamics of the system under investigation. For particles of size of order microns, choosing visible wavelengths makes the scattering efficient and strongly dependent on particle properties such as size. However the stronger the interaction the more likely that multiple scattering will be a problem. If each photon interacts with only one particle (i.e. single scattering) it arrives at the detector carrying unambiguous information about that particle. If on average it interacts with more than one particle (multiple scattering) the information it carries relates to all of these particles and may be ambiguous or difficult to interpret. Most popular techniques assume single scattering. As the particle concentration increases, at some level this assumption will become inadequate. This paper is concerned with estimating the error due to multiple scattering in planar Mie images of sprays.

In planar Mie imaging a laser beam is formed into a thin sheet which illuminates a plane in the particulate field. A camera, axis perpendicular to the light sheet, images the scattered light. The image yields information about the spatial distribution of particles in the illuminated plane. If the particle size is known the concentration can be quantified. The optical setup of planar Mie imaging is the basis of other imaging techniques which yield other quantities, e.g. planar laser induced fluorescence.

Multiple scattering affects planar imaging in several ways. Scattering near the light sheet broadens the sheet, reducing spatial resolution in the camera axis direction. Scattering outside the light sheet reduces the contrast and clarity of the image. These effects depend on the three dimensional structure of the particle field.

It is difficult to assess the effects of multiple scattering experimentally. A common rule of thumb is that if the line of sight transmittance is less than 90% then multiple scattering is significant (e.g. Jones 1993), but this tells us nothing about the effects on the image. Information about the number of scatters experienced by a given photon is encoded in its properties of direction, position, polarisation and time of arrival but in an ambiguous manner. It is more convenient to study the effects computationally.

To reproduce the effects of multiple scattering in a calculation the radiation transport equation must be solved. Analytic solutions are rarely possible for practical geometries. The diffusion approximation is not valid at the intermediate concentrations in which planar imaging is applied. The equation must be tackled by discretisation and numerical solution, either deterministic or stochastic (Monte Carlo). The Monte Carlo approach was chosen for this work because it is simple to implement and the calculations can be stopped at any time to inspect the results, and continued if more accuracy is required.

The Monte Carlo approach has a long history in scattering calculations in medical diagnostics (e.g. Key et al. 1991, Hiraoka et al. 1993, Wang et al. 1995, Tuchin 2000, Meglinsky and Matcher 2001) and meteorology (e.g. Marchuk et al.. 1980 and Kokhanovsky, 2001). It has been used very little to analyse the performance of optical diagnostics for flow systems. Giddings et al.. (2000) used Monte Carlo calculations to analyse the error due to multiple scattering in planar Mie images of a two-gas mixing experiment, in which one gas was seeded with oil droplets. The authors have used the code described in this paper to analyse planar Mie images of sprays (Jermy and Allen, 2002).

2. The Monte Carlo code SATURN

The code is written specifically to handle the highly three-dimensional geometries encountered in sprays. The space occupied by the spray is divided into small cubic cells. The droplet number density is specified in each cell. Some cells are defined as light sources and given a direction and a strength. The code loops through all source cells, releasing from each source cell a number of photons proportional to its strength. Each photon is tracked on its journey through the spray as follows. As it enters each cell, the distance it would travel before leaving the cell, if no scatter occurs, is calculated. The probability of scatter is then calculated as the product of this path length in the cell, the droplet scattering cross section, and the number density of droplets in the cell. A pseudorandom number is generated and compared to this probability. If the number is greater than the probability of scatter, no scatter occurs and the photon moves on to the next cell. If scatter does occur, the photon is assumed to scatter a random point on the original trajectory within the cell, and given a new direction. Two pseudorandom numbers are generated and used to select two angles from the scattering phase function. The photon direction is deviated by these angles. It is then moved to next cell. The calculation proceeds until the photon leaves the modelled space. Its direction, position and number of scatters are recorded and the next photon is started.

A postprocessing code models the camera. It is fed with information on the position of the camera and the lens object plane, the lens aperture, and the size of the pixel array in the image plane. The camera code processes each photon leaving the modelled space as follows. It projects the photon trajectory to the point at which it intersects the plane of the lens and determines if this point lies within the lens aperture. If so, the photon contributes to one pixel in the image. Which pixel benefits is determined by finding the point at which the photon trajectory intersects with the objects plane. This point lies within the projection (onto the object plane) of one of the pixels. A tally is kept for each pixel for each order of scatter. First order photons are those that have been scattered only once, second order those that have been scattered twice and so on. The code can produce a simulated image from all photons (all orders), equivalent to that produced by a real camera which cannot distinguish between photons that have been scattered once and those that have been scattered multiple times. The code can also produce an image from first-order photons only, i.e. a perfect single-scattering image, or from second-order photons, and so on. Comparing such images reveal the effects of multiple scattering.

3. Simulations of a hollow-cone spray

Planar Mie imaging of a hollow cone spray was simulated. The spray had characteristics similar to those of the pressure-swirl sprays used in oil fired furnaces and gasoline direction injection. A 50mm cubic volume was modelled, divided into cubes of 1mm side. The droplets were all 20μm in size with a droplet number density n given by Eq 1 which yields a hollow cone with an included angle of 90°, a peak number density $\approx$100/mm^3 and a peak packing fraction of 0.04%.

$$n = \begin{cases} \dfrac{100r^2}{z^2} & r \leq z \\ 0 & r > z \end{cases} \qquad \text{Eq 1}$$

The light sheet had no divergence, was 1mm thick, 50mm high and included the axis of the spray. The wavelength was 532nm. The medium had refractive index 1, the droplets 1.4 and neither were absorbing. The phase function was calculated with Michel's Mie code. 2×10^8 photons were tracked.

The lens had aperture 60mm and was placed 75mm from the light sheet, on which it was focussed. The camera had 30x30 pixels.

To determine in which parts of the spray the worst multiple scattering occurs, the simulation was repeated three times: once with the full spray cone, once with the spray outside the light sheet on the camera side removed, and once with all spray removed except that lying in the path of the light sheet (Fig. 1).

Fig. 2 shows the simulated images for all, first and second order photons. Fig. 3 shows plots of the photon count from the 11th row from the bottom. From Fig. 2 it is clear that the light sheet is attenuated as it travels through the spray. This attenuation has been corrected in Fig. 3 by the method of Talley et al. (1996). Also plotted in Fig. 3 are the droplet number density, which would be identical in shape to the intensity profile of the image if single scattering held and the image reconstruction were perfect. Density is presented on two scales to ease comparison of to the intensity field of the full cone (most imperfect) and plane (most perfect) images.

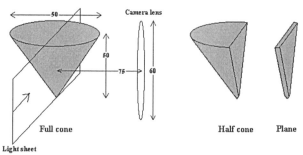

Figure 1: Three geometries used for the hollow cone spray simulations, Dimensions in mm.

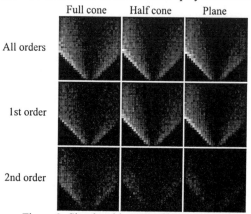

Figure 2: Simulated images of hollow cone spray

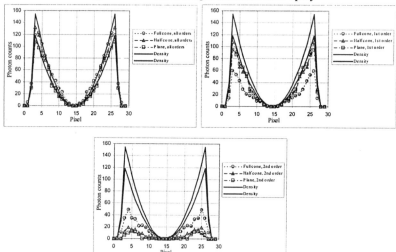

Figure 3: Photon count and density profiles from hollow cone spray simualtions

4. Discussion

In the full-cone simulation, multiple scattering occurs at significant levels- up to 50% of the photons reaching the camera have been scattered more than once. However in the half-cone and plane simulations the levels of multiple scattering are much lower. The half cone and plane profiles are very similar. The illumination is symmetrical so if multiple scattering occurs in one half of the spray it must occur in the other. Therefore the multiple scattering *that contributes to the image* is dominated by that which occurs in the half of the spray that lies between the illumination plane and the camera.

There is no significant blur in the all-orders images or profiles, despite the high levels of multiple scattering. In these simulations the scattering is forward dominated: secondary and subsequent scattering events tend to deviate the photon trajectory by very small amounts. If the particles have significant absorbance (as in the case of sprays doped for laser induced fluorescence) or the scattering is not so strongly forward-scatter dominated (smaller particles or longer wavelengths) the effects of multiple scattering will be worse. The image will suffer attenuation with a spatial profile of the same shape as the higher scattering orders in Fig. 3.

For the full cone there appears to be a slight reduction in intensity at the very edges of the spray, perhaps due to the long path length through the spray from these points to the lens. This reduction is not apparent in the half-cone simulation. Again, the part of the spray that lies between the illumination plane and the camera is distorting the image.

4.1. Effect of camera aperture

The effect of the camera aperture on the amount of multiply scattered photons reaching the image was explored by processing the results of the full cone simulation with the camera at 125mm from the illuminated plane and with aperture radii of 2.5, 5, 10, 15, 20, 25, 30, 40, 45, 50, 75, 100 and 150mm. Figure 4 plots the total number of singly scattered divided by the total number of photons (of any order). The ratio for doubly scattered photons is also plotted. The error bars are derived from the shot noise (square root of the photon count) and include contributions from the shot noise in the 1st or second order count and in the total count.

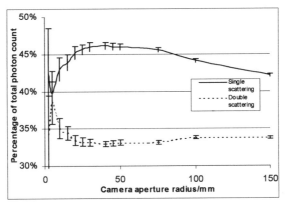

Figure 4: Percentage of single and double scattering reaching image as a function of aperture.

Fig. 4 shows that the contribution of multiply scattered photons to the image falls off as the aperture is increased from the smallest values. As the aperture is increased further, becoming larger than the object, the contribution of multiple scattering increases again, slowly. This may be due to the multiply scattered photons travelling on average at higher angles to the camera axis than the singly scattered. The camera is closer to the spray than is typical for real imaging experiments, so this latter effect may not be important in realistic geometries. In any case changes in camera aperture account for a change in the contribution of multiple scattering of only $\approx 7\%$.

5. Conclusions

A flexible photon transport code has been developed, capable of simulating arbitrary sources, three dimensional scattering fields of arbitrary shape, and many scattering mechanisms. For two phase flow imaging and other turbid, particulate or colloidal media problems, the code allows the amount and effects of multiple scattering to be quantified, and corrective strategies to be explored.

It has been applied to the type of planar Mie imaging situations common in two-phase flow studies. In such situations some tens of percent of the photons reaching the camera may be multiply scattered. Under the conditions modelled, changing the camera aperture is ineffective in reducing the contribution of multiple scattering to the image. The particles modelled here are non-absorbing, Mie-scattering and significantly larger than the wavelength of the illumination. Their scattering is forward dominated and this explains two of the results. The multiple scattering that affects the image originates in the half of the spray that lies between the illuminated plane and the camera. Also, multiple scattering causes little blur or position-dependent attenuation in the images, due to the majority of secondary scattering events being forward-scattering and thus deviating the photon little. For particles nearer the Rayleigh limit, forward scattering is not so dominant and blur may be more serious. For thicker systems and systems with absorbing particles the effects are also expected to be more serious.

References

Giddings R, Holder DA, Philpott M and Youngs DA 2000, "Multiple scattering from particles in laser sheet gas mixing experiments", *Advances in Statistical Optics*- IOP half day meeting, Imperial College, London (24th May 2000)

Hiraoka M, Fibrank M, Essenpreis M, Cope M, Arridge SR, van der Zee P and Delpy DT 1993, *Phys. Med. Biol.* **38** 1859-1876

Jermy MC and Allen A 2002, *Appl. Opt.* **41** (in press)

Jones AR 1993, *Light Scattering for Particle Characterisation*, in *Instrumentation for Flows with Combustion* (Ed. A.M.K.P. Taylor, Academic Press., London), pp. 323-404

Key H, Davies ER, Jackson PC and Wells PNT 1991, *Phys. Med. Biol.* **36** 591-602

Kokhanovsky AA 2001, *Optics of Light Scattering Media,* Springer-Verlag, Berlin

Marchuk GI, Mikhailov GA, Nazaraliev MA, Darbinjan RA, Kargin BA, Elepov BS 1980, *The Monte Carlo method in atmospheric optics,* Springer, Berlin

Meglinsky IV and Matcher SJ 2001, *Med. Biol. Eng. Comput.* **39** 44-50

Michel B, http://www.unternehmen.com/Bernhard-Michel/Java/mieApplet.html

Talley DG, Verdieck JF, Lee SW, McDonell VG and Samuelsen GS 1996, AIAA 96-0469

Tuchin V 2000, *Tissue Optics*, SPIE Press

Wang L, Jacques SL and Zheng L 1995, *Comput. Methods Programs Biomed.* **47** 131-146

Inst. Phys. Conf. Ser. No. 177
Paper presented at 1st Int. Conf. on Optical & Laser Diagnostics, London, 16–20 Dec. 2002

Optical Diagnostics of Pressure and Temperature using Laser Induced Thermal Gratings

R Stevens and P Ewart

Oxford Institute for Laser Science, Physics Department, Oxford University, Oxford, UK

Abstract. Laser induced thermal grating spectroscopy (LITGS) is based on scattering from gratings produced by interference in a resonantly absorbing gas, of two pump laser beams crossing at a small angle. Interference between a stationary temperature grating and induced acoustic waves leads to a modulation of the grating scattering efficiency. The dynamics of the grating evolution and decay may be monitored by scattering of a probe beam incident at the appropriate Bragg angle.

Previous studies have used cw probe lasers with power of approximately 1 W and so the scattered signals are limited by the energy incident during a grating lifetime of typically 0.1 to 1.0 μsec. In the present work we demonstrate the use of long pulse probes (pulse duration 1.5 μsec) with powers of up to 1 MW yielding a potential dramatic increase in detection sensitivity. LITGS in NO_2 in buffer gas pressures from 1 to 40 bar are detected in concentrations down to 100 ppm. Simulations, based on linearized hydrodynamic equations, fitted to the data allow accurate time and space resolved measurements of temperature and pressure. Applications to other species of combustion interest, flame and engine measurements will be discussed.

Inst. Phys. Conf. Ser. No. 177
Paper presented at 1st Int. Conf. on Optical & Laser Diagnostics, London, 16–20 Dec. 2002
©2003 IOP Publishing Ltd

Fibre Optical Diagnostics for structural condition monitoring applications exemplified in the MILLENNIUM project

Y M Gebremichael, W Li, B T Meggitt[+], W J O Boyle, K T V Grattan, B McKinley & LF Boswell

School of Engineering & Mathematical Sciences, City University, Northampton Square, London EC1V 0HB, UK
[+]EM Technology, 8 Whitfield Court, London SE21 8AG, UK

Abstract. The MILLENNIUM project is a recently completed EU-funded exercise, a major part of which was to instrument a Norwegian bridge to achieve better monitoring of the structure in use. To do so an effective, thermally-compensated strain measurement technique using a series of strain-isolated fibre Bragg gratings has been demonstrated. An important aspect of the project was the field testing of the multi-channel fibre Bragg grating sensor system, developed as part of a new condition maintenance programme. This enables real time online integrity monitoring for not only the new Norwegian steel road bridge but potentially for other civil engineering structures. The results of laboratory and field trials are presented as are further applications of the technology in the civil engineering sector.

Keywords: Temperature and strain discrimination, temperature compensation, Bragg grating sensors

1. Introduction

Fibre optic sensors have shown a considerable potential to meet a wide range of measurement needs in industry and commerce, but it is only in recent years that several important practical applications have been demonstrated outside the laboratory. The interest in the technology arises in part as a result of their inherent nature in being light in weight, their immunity to electro-magnetic interference, their showing damp/corrosion resistance and offering potentially high measurement sensitivity. Various parameters including strain, temperature, pressure, force, displacement, vibration, film thickness, refractive index, and pH etc have been measured with this type of sensor, as discussed in detail elsewhere [1]. A key application area is in civil engineering structures, as the fibres are compatible with metals, composites and concrete. A disadvantage that comes with this type of measurement is that a cross sensitivity to several of the above parameters often exists within a practical fibre optic sensor system, and this issue is addressed and discussed in this work in the design of the sensor system deployed.

2. The MILLENNIUM Project

The EU-funded project under the Brite/EURAM initiative involved a number of industrial and partners with complementary experience to engage in both the design, testing, installation and evaluation of a novel multi-channel fibre optic strain monitoring system, coupled with appropriate modelling at both the physical level (on a 10 m scale model) and through advanced computational techniques. The focus of this paper is to address these design aspects of a sensor system, developed for a specific application in strain monitoring, and to deal with temperature/strain cross sensitivity and the consequent achievement of thermal compensation. The system utilized a fibre optic Bragg grating sensor system designed for long term monitoring and results are reported on the system design and performance.

This approach has been applied on the field structure considered in the MILLENNIUM project, the Mjosund steel Road Bridge in Norway. The work was designed to support the Norwegian Roads Authority to monitor better the many bridges joining the island coastline of Norway to the mainland. The bridge was a 346m long steel box section structure with a concrete platform (Figure 1) carrying the road-access to the bridge, and access for the fibres and test equipment was through the box section. The ~2m height enabled comfortable human access for installation of the equipment for long term testing, over more than a year and with that the significant temperature changes from summer to winter in Norway.

As will be shown in this work, for structural integrity monitoring and fatigue analysis of existing large structures such as this bridge, the aim is for *specially identified sections* of the structure to be instrumented with a cluster of sensors per section, to give the optimum information on the performance of the structure, following the lead of the civil engineers responsible for maintenance. Fortunately, in such applications a single strain-isolated temperature sensor per section is usually sufficient for thermally compensated measurement, thus saving on the total number of sensors needed

Figure 1 Mjosund bridge, Norway, 346m long

3. Fibre Optic Bragg grating-based sensors

In this work the decision was made to use fibre Bragg gratings (FBGs) as the essential sensor component. The theoretical background to the use of such systems is discussed elsewhere [1] but in summary a FBG arises from a periodic perturbation of the refractive index of the core of the optical fibre written usually in a single mode fibre with a UV light interference pattern. The refractive index modulation over a few millimetres of the fibre makes the Bragg grating essentially a wavelength selective mirror, such that the grating reflects a narrow band of the incoming broadband radiation [2]. This reflected spectrum of the grating is characterised by the Bragg equation, $\lambda_B = 2n\Lambda$, where λ_B is the centre wavelength of the reflected spectrum, n is the core refractive index and Λ is the pitch length. The sensor response arises from changes in both the grating pitch length and the perturbation of the effective core refractive index, which translate into a wavelength shift. Such changes can be induced due to variations in strain, temperature, pressure etc. Strain and temperature measurements are of particular interest in Bragg grating sensor applications for long-term structural integrity and condition monitoring of large civil or industrial structures. In this work, such FBGs are subjected to mechanical strain and with that an inevitable change in the ambient temperature. The combined strain and temperature sensor response (given by a single combined Bragg wavelength shift, $\Delta\lambda_B$) can be represented by the linear relationship:

$$\frac{\Delta\lambda_B}{\lambda_B} = (\alpha + \xi)\Delta T + (1 - \rho_e)\Delta\varepsilon \qquad (1)$$

where $\Delta\varepsilon$ is change in strain, ΔT is the change in temperature, α ($=0.55 \times 10^{-6}/^\circ C$) is the fibre linear thermal coefficient and ξ ($= 8.3\times10^{-6}/^\circ C$) is the thermo-optic coefficient. The first two terms in equation (1) above represent effects due to the temperature variation on the grating. It can be shown that for a strain-isolated fibre grating, the wavelength shift with temperature is mainly due to the change in the core refractive index and not simply the fibre linear expansion [2]. For a free grating at 1550nm, $\Delta\lambda_B/\Delta T = 13.7$pm/$^\circ$C. The second part of equation (1) reflects the effect of strain on the Bragg wavelength shift, where the strain-optic coefficient, ρ_e is given by $n^2/2(\rho_{12} - v(\rho_{11}+\rho_{12}))$, where n is the core refractive index, v ($= 0.16$) is the Poisson ratio and ρ_{11}, ρ_{12} ($=0.252, 0.113$ respectively) are the components of the strain optic tensor. Thus $\rho_e = 0.26$ and ($\Delta\lambda_B/\lambda_B$)$=0.74\Delta\varepsilon$ and for a free grating at 1550nm, and $\Delta\lambda_B/\Delta\varepsilon = 1.15$pm/$\mu\varepsilon$. The above equation illustrates the problem of cross sensitivity of temperature and strain, which together results in the wavelength shift that is measured. Their deconvolution has been the subject of considerable research work in recent years and several techniques have been published to address the issue of temperature and strain discrimination in Bragg based sensors. One approach employs the use of two superimposed fibre Bragg gratings at two different wavelengths, λ_2 and λ_2 [3] from which the temperature and strain components of the measured wavelength shift can be discriminated simultaneously using two linear equations. Another technique involves the use of a fibre Fabry-Perot (F-P) cavity and a fibre Bragg grating in which the F-P cavity senses strain and the Bragg grating measures both the strain and temperature [4]. The use of fibre Bragg gratings with different claddings spliced together, when subjected to strain, results in different levels of strain being experienced, these relating to their differing physical diameters while their temperature response remains similar. A differential measurement allows decoupling of these two parameters [5]. Similarly a grating written on the splice joint between two different fibres with the same diameter but having differing refractive indices produces two peaks. Owing to their equal diameters, the two fibres experience a similar variation in the grating pitch for a given load, while the refractive index variation equates to a differing response to temperature change in each. Consequently the temperature and strain responses can be discriminated [6].

Alternatively two gratings in series, where one is isolated from the strain and a second Bragg sensor (to measure both strain and temperature) have been used to decouple strain and temperature [7]. Another example is a hybrid sensor in which both a long period grating (LPG) and an FBG are used [8]. Additional methods include a Bragg grating written on a fluorescent fibre which has also been employed for measuring temperature and strain by combined decay time and wavelength shift measurement [9]. Further, a temperature independent strain sensor using a chirped grating, partially embedded in a glass tube has been reported [10]. Analogous techniques exist involving various interferometric and polarimetric techniques for strain and temperature discrimination and these are discussed elsewhere [11].

4. Measurement effects and sensor discrimination

4.1 Temperature compensated measurement

In structural health monitoring, the Bragg grating sensor is normally either surface-bonded to the structure or embedded within it. In either case, the Bragg sensor responds to strain and temperature changes together giving a combined wavelength shift, which yields a measurand termed here the *apparent strain*. This is composed of three components, namely the mechanical loading strain, the thermally induced strain due to the thermal expansion and contraction of the structure and a third component due to the thermo-optic effect on the fibre itself. While in some applications it may be useful the discriminate the thermal strain from the loading strain, for structural integrity and fatigue analysis, both the loading strain and the strain due to the thermal response of the structure are important parameters in determining performance failure. In such cases, only the strain component due to the thermo-optic effect on the fibre and the small thermal expansion of the fibre needs to be taken in to consideration for effective thermal compensation to be obtained.

4.2 Strain and temperature discrimination

Two approaches based on strain isolated Bragg grating sensor are discussed in this work, in relation to their applicability to the "real world" measurement problems illustrated here. The simplest and most direct technique is a differential method, where differential measurements are taken between a fibre strain gauge attached to the structure and a second strain isolated sensor grating, which is loosely placed on the structure, without it being affected by the inevitable thermal expansion or mechanical loading of the structure. The differential strain provides a combined loading strain and strain induced by the thermal expansion of the structure. To demonstrate this effect in the laboratory tests prior to the installation on the bridge, three gratings were placed inside a stable test oven. One of the gratings was bonded to a steel bar, a second grating to a silica rod and a third grating was loosely placed on the steel bar near the bonded grating so that it was free from strains (due to mechanical or thermal loading of the bar) but it experiences the temperature effects on the fibre. The oven temperature was programmed to ramp from room temperature to 75°C in steps of 10°C and measurements were taken when the oven temperature was stabilised at a set point. At each temperature set point, a fixed load is applied to the beam so that the loading strain can also be recorded. Figure 2 below shows the measurement results obtained from varying the temperature of the oven from room temperature to 75°C. Trace 'A' shows the wavelength shift of an isolated fibre with temperature, where the slope of the curve is 10pm/°C, while Trace 'B' represents the wavelength shift for a grating attached to a silica rod subjected to the same temperature change as the isolated grating. The thermal expansion coefficient of silica is very close to that of the fibre, and thus the wavelength shift recorded is close to that of the isolated fibre. Traces 'C' and 'D' represent data from the grating bonded to the steel beam, for unloaded and loaded situations respectively. Here $\Delta\lambda/\Delta T$ is recorded to be 26pm/°C and load is applied at 30°C.

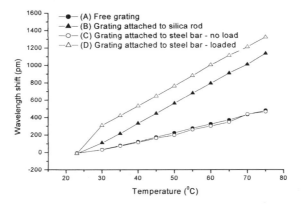

Figure 2 Wavelength shift with temperature for free and attached FBGs under different loading circumstances

Applying the differential technique to such measurement data, temperature and strain can be decoupled as shown in Figure 3. In this experiment, a load is applied to the steel cantilever when the oven temperature is recorded as 35°C, as denoted by the step jump in the traces at this temperature. The component of the wavelength shift due to thermo-optic effect on the fibre is compensated by subtracting the strain-free data, measured directly from the isolated grating (trace A) and from the strain sensor data, measured by the gratings bonded to the structure (trace B). The resulting compensated data (trace C) gives the structural strain due to the thermal expansion of the structure plus the measured loading strain. It is also possible to discriminate between thermal strain and mechanical loading strain by subtracting the thermal slope of the strain. The resulting strain is shown as trace D in Figure 3 below.

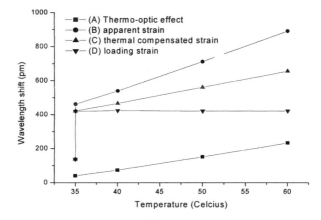

Figure 3 Apparent strain and compensated strain using the differential technique

In field applications, however, a loosely attached fibre cannot guarantee a long-term thermal contact with the structure and may thus give a false temperature reading. In this regard it is necessary to attach the strain-isolated sensor to the structure to provide the required structural temperature data. The response of the attached strain-isolated sensor has been shown to be similar to that of a strain-free fibre, carefully placed in thermal contact with the structure in the laboratory experiments where the FBG sensor is attached to a silica rod which showed the same $\Delta\lambda/\Delta T$ gradient, as is shown in Figure 2.

In large structures, such as the Mjosund steel Road Bridge for which the monitoring system has been designed, it was found impossible to find strain-free sections for all loading conditions. Hence a small piece of steel was attached to the structure so that it did not experience the mechanical loading strain, as shown in Figure 4. It should be noted that the sensor attached to the structure is constrained by the structural boundary conditions, while the sensor attached to the steel piece is free to expand with temperature. Thus subtracting the apparent strain readings of the isolated sensor from those for the sensor on the large structure will not represent the mechanical loading, but a resulting value, which is less than that which the structure actually experiences. Furthermore, for structural health monitoring, the strain induced on the structure due to the thermal loading is as important as the mechanical loading strain. Thus, although it may be helpful to identify the components of the combined strain, it should be noted that taking a simple numerical difference (subtracting out) the thermal effects, gives a false strain reading. The only component that should be compensated is the strain reading due to fibre thermo-optic effect and the small thermal expansion of the fibre.

Figure 4 Photograph showing strain isolated temperature sensor on a 10m steel box model bridge in the laboratory

5. Measurements made on the Mjosund steel Road Bridge in Norway

The laboratory tests and analysis carried out were applied in an actual structural analysis situation. Thus the wavelength shift measured with the gratings bonded to the strain-isolated steel-piece, which in turn is attached to the structure for thermal contact, arises due to the thermo-optic effect on the fibre itself and the strain induced on the grating with the expansion/compression of the steel due to the change in temperature, as shown by equation (2) below.

$$\Delta\lambda = \lambda\left[\left(\xi + \alpha\right)_{Fibe} + \left(1 - p_e\right)\alpha_{Steel}\right]\Delta T \tag{2}$$

where α_{Steel} is the thermal expansion coefficient of steel. Knowledge of the thermal expansion of the steel can be used with the wavelength shift data, $\Delta\lambda$, measured from the strain isolated grating attached to the steel-piece to extract the temperature of the structure and use that value to correct for the contribution of strain arising from the thermo-optic effect of the fibre.

A grating attached to the steel in the oven, monitored with varying temperature, recorded a value of $\Delta\lambda/\Delta T$ of 26pm/°C and thus $\Delta T/\Delta\lambda$ was determined to be 38.5°C/nm. Multiplying the wavelength shift, $\Delta\lambda$ measured by use of the temperature compensation grating (the strain-isolated attached sensor) by $\Delta T/\Delta\lambda$ gives the structural temperature variation, ΔT, from which the Bragg wavelength shift due to the thermo-optic effects can be extracted. This is done by multiplying the inferred structural temperature with the wavelength shift per °C factor of a free grating, from which the equivalent strain due to the fibre thermo-optic effect could be extracted. Sensor gratings attached to the steel structure were used to measure wavelength shifts due to thermal as well as loading strain. Subtracting the resulting measurement of strain due to the thermo-optic effect from the apparent strain measured by using the sensor gratings gives the thermally compensated strain. Figure 5 shows experimental data from both an FBG sensor and (for comparison) a resistive strain gauge attached to a steel cantilever placed inside an oven, to demonstrate the thermal compensation technique.

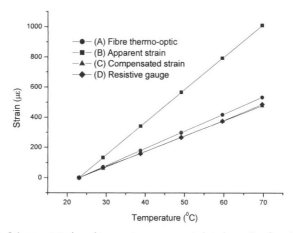

Figure 5 Apparent strain and temperature compensated strain reading from FBG sensor and equivalent strain reading with a resistive strain gauge

The data obtained show that the resultant thermally-compensated strain measurements closely agree with the strain measured by using the electrical strain gauge, which itself is inherently compensated for temperature.

A key feature of the use of the Bragg grating technique is the development of a series of optical sensors multiplexed along a single optical channel. For multiple measurement points, the individual connection of what may be several hundred strain gauges becomes a major logistical exercise – the fibre sensors were arranged in several different channels with 4 or 5 grating sensors per channel providing a total of 32 measurement points with 8 parallel channels. Figures 6 and 7 show representative comparative measurement data between 29 FBG sensors and 8 resistive gauges installed at various predefined (by the civil engineers) points on the Mjosund Road Bridge, which in total had 32 FBG sensors and 12 conventional strain gauges along with thermocouples. The sensor configuration was such that for each resistive foil gauge, there were two FBG sensors placed at the same strain point for data verification. Different loading conditions were applied at different times over the test period of 14 months (thus different temperatures). From the study carried out, application of the thermal compensation technique discussed in this paper shows that the resulting thermally-compensated strain data obtained optically closely matches the readings from the resistive gauge data to within less than 5με (equivalent to the system noise) as shown in Figures 6 and 7. It should be noted that the temperature variation on the location of this particular bridge ranges from -40° in the winter to +25°C in the summer, which on its own was shown responsible for the most part of the strain the structure and the bonded fibre sensors experienced, further highlighting the importance of thermal compensation in such applications.

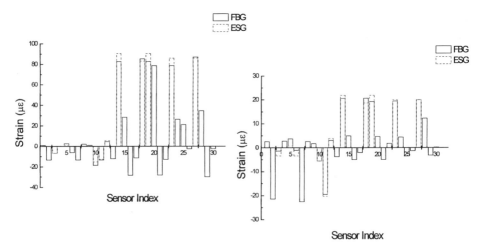

Figure 6 & Figure 7 Strain measured at various points across the bridge during loading – a comparison of fibre sensors (FBG) and electrical strain gauge (ESG) results over 18 months

6. Discussion

Data have been presented demonstrating the feasibility of using a mechanically decoupled FBG for temperature compensated strain measurements, illustrated in a practical civil engineering application. It has been shown that in uses such as in structural health monitoring of large civil engineering structures, the main issue in thermal compensation using these optical devices is the thermo-optic response of the sensor grating with temperature. While it is true that the thermal loading of the structure gives apparent strain readings, in many civil engineering applications this measurement is as important as the mechanical loading of the bridge due to heavy traffic. Thus when thermal compensation using FBG methods in such applications is discussed, it should be noted that reference needs to be made to the thermo-optic effect of the fibre due to the temperature dependence of the fibre refractive index. For applications such as to the Mjosund steel Road Bridge under study here, where long term structural health and fatigue analysis is important, it has been shown in this work that thermal compensation by use of a mechanically decoupled temperature sensor is a practical solution in experimental measurements.

Acknowledgements

The authors are pleased to acknowledge support from the EU sponsored Brite EURAM 'Millennium' project on the monitoring of large civil engineering structures for improved maintenance, the Engineering and Physical Sciences Research Council (EPSRC) structural integrity programme and other EPSRC initiatives.

References

[1] KTV Grattan & BT Meggitt " Optical Fibre Sensor Technology: Fundamentals" Kluwer Academic Press, 2000.
[2] A Orthonos " Bragg grating in optical fibres: Fundamentals and applications " in Optical Fibre Sensor Technology: Advanced Applications (Eds KTV Grattan & BT Meggitt), Kluwer Academic Press, 2000, p.79 – 188.
[3] MG Xu, JL Archambault, L Reekie & JP Dakin " Discrimination between strain and temperature effects using dual wavelength fibre grating sensors" Electronics Letters (1994) Vol 30 No. 13.
[4] T Liu, GF Fernando, L Zhang, I Bennion, YJ Rao & DA Jackson "Simultaneous strain and temperature measurement using a combined fibre Bragg grating/extrinsic Fabry-Perot Sensor" 12th International Conference on Optical Fibre Sensors (1997) Vol. 16 p.40 – 43.
[5] SW James, ML Dockney & RP Tatam " Simultaneous independent temperature and strain measurement using in-fibre Bragg grating sensors" Electronics Letters (1996) Vol. 32 No. 12.
[6] B Guan, H Tam, S Ho, WH Chung and XY Dong "Simultaneous strain and temperature measurement using a single fibre Bragg grating" Electronics Letters (2000) Vol. 36 No. 12.
[7] F M Haran, JK Rew & PD Foote " A strain – isolated fibre Bragg grating sensor for temperature compensation of fibre Bragg grating strain sensors" Meas. Sci. Technol. 9 (1998) p. 1163 – 1166.
[8] HP Patrick, GM Williams, AD Kersey, JR Pedrazzani & AM Vengsarkar. "Hybrid fibre Bragg grating/long period grating sensor for strain/temperature discrimination." IEEE Photon. Technol. Lett. (1996) Vol 8 p 1223 – 1225.
[9] DI Forsyth, SA. Wade, T Sun, X Chen and KTV Grattan, "Dual temperature and strain measurement using the combined fluorescence lifetime and Bragg wavelength shift approach in doped optical fiber", Applied Optics (Accepted for publication, 2002).

[10] S Kim, J Kwon, S Kim & B Lee, " Temperature independent strain sensor using a chirped grating partially embedded in a glass tube " IEEE Photon. Technol. Lett. (2000) Vol. 12, No. 6.

[11] JDC Jones, "Review of fibre sensor techniques for temperature-strain discrimination" 12th International Conference on Optical Fiber Sensors, (1997) Vol. 16 p.36 – 39.

[12] YM Gebremichael, BT Meggitt, WJO Boyle, W Li, KTV Grattan, L Boswell, B McKinley, KA Aarnes & L Kvenild. " Multiplexed fibre Bragg grating sensor system for structural integrity monitoring in large Civil Engineering applications" Proceedings of the 11th Conference on Sensors and Applications XI, London (2001) p. 341 – 345. (Pub. Institute of Physics, London, UK: Eds KTV Grattan and SH Khan).

Inst. Phys. Conf. Ser. No. 177
Paper presented at 1st Int. Conf. on Optical & Laser Diagnostics, London, 16–20 Dec. 2002
©2003 IOP Publishing Ltd

Quantitative NO-LIF imaging in high-pressure flames

W G Bessler*, C Schulz*, T Lee, D-I Shin, M Hofmann,
J B Jeffries, J Wolfrum*, and R K Hanson

Mechanical Engineering Department, Stanford University, Stanford CA 94305

* Physikalisch-Chemisches Institut, Universität Heidelberg, 69120 Heidelberg

Abstract. Quantitative laser-based imaging of NO concentrations is important for many practical high-pressure combustion applications. With increasing pressure, however, it is increasingly affected by interference, laser- and signal absorption and temperature variations. In steady laminar flames multiple sequential laser-based measurements can provide corrections to increase the over-all accuracy; however, in practical combustors with turbulent flames not all required correction data can be obtained simultaneously on a single-shot basis. We quantitatively investigate the influence of various corrections on NO laser-induced fluorescence using detailed measurements in laminar methane/air flat flames at $1 - 60$ bar. We investigate the influence of O_2 interference, the dependence on local temperature and the effect of laser and signal attenuation. Despite using a NO detection scheme with minimum O_2-LIF contribution, the fluorescence interference yields errors of up to 25% in the slightly lean 60bar flame. The over-all dependence of NO number density on temperature in the relevant range is low ($< 6\%$ for a 200 K temperature variation) because different effects cancel. In contrast, attenuation of laser and signal light by combustion products CO_2 and H_2O, which is usually neglected, yields errors up to 40% despite the small scale (8 mm diameter) of our experiment. In practical devices, these attenuation effects may be the major source of errors. Understanding the dynamic range for each of these corrections provides guidance to the uncertainty in single shot images at high pressure.

1. Introduction

Quantitative imaging of nitric oxide (NO) concentrations with laser-induced fluorescence (LIF) has attracted significant interest in recent years for optimizing combustion efficiency and minimizing pollutant formation as well as for validating combustion chemistry models [1-5]. At high pressures, however, NO-LIF spectroscopy is faced with a number of problems that complicate quantification. Fluorescence interference occurs from O_2-LIF in high-pressure high-temperature fuel-lean environments [6]. The interpretation of NO-LIF signals is influenced by temperature in many respects, since not only ground state population distribution, but also line shapes and fluorescence-quenching cross-sections are temperature-dependent. The quenching rates depend on local gas composition, which also varies with temperature. In the < 250 nm range used for NO diagnostics, strong absorption of laser and signal light by combustion products such as H_2O and CO_2 must be considered [7], and this absorption strongly depends on temperature. Finally reliable calibration techniques are required [2,8]. These influences are discussed in more detail in a related paper from the same authors [9].

In steady laminar flames these potential errors can be accounted for by sequentially measuring NO-LIF, interfering background-LIF intensities, and temperature and modeling concentration distributions of major species. We assess here the relative importance of the different corrections in steady, laminar methane/air flames at 1 – 60 bar in order to determine the error margins introduced when these corrections are not quantitatively known (e. g. in the investigation of turbulent flames commonly found in practical combustors).

2. Experimental

Laminar, premixed methane/air flat-flames at pressures from 1 – 60 bar were stabilized on a porous, sintered stainless steel plate of 8 mm diameter. Investigations were conducted for $\phi = 0.95$ fuel/air equivalence ratio. An additional 2% (by volume) NO/N_2 mixture with variableNO concentration was added to the feedstock gases to allow NO seeding at the 0 – 600 ppm level without changing the dilution of the flame gases. The total flow rates are listed in table 1. Laser light (2 mJ @ 224 – 227 nm, 0.4 cm^{-1} fwhm) from a Nd:YAG-pumped (Quanta Ray GCR250) frequency-doubled (BBO) dye laser (LAS, LDL205) was formed to a vertical light sheet (5 x 0.5 mm^2) crossing the flame horizontally. Fluorescence signals were collected at right angles to the laser beam and focused with a $f = 105$ mm, $f/4.5$ achromatic UV-lens (Nikon) onto the chip of an intensified CCD camera (LaVision FlameStar III). The signal light (A-X(0,1) around 237 nm) was observed through a combination of dielectric mirrors. The experimental setup has been described in detail elsewhere [9].

The laser was tuned to the NO $A\,^2\Sigma^+$–$X\,^2\Pi(0,0)$ P$_1$(23.5), Q$_1$+P$_{21}$(14.5), Q$_2$+R$_{12}$(20.5) absorption feature at 226.03 nm used previously by DiRosa et al.[6]. This excitation wavelengths minimizes O_2-LIF interference [10]. The remaining background was measured off the NO resonance at 227.48 nm. This wavelength was chosen because it yields a comparable intensity of O_2-LIF like 226.03-nm excitation. The images averaged over 50 laser shots were corrected for the light sheet intensity variations and spatial variations in detection efficiency.

The local temperature at each pixel was determined from a multi-point fit of the NO-LIF excitation spectra [11] in the 225.95 – 226.10 nm range for pressures up to 40 bar with 300 ppm NO seed. This region corresponds to a broad minimum in O_2-LIF interference [6]. The laser is scanned (in 0.001-nm increments) and a stack of 2d-images is obtained. The resulting three-dimensional data base (two spatial coordinates and the excitation wavelength coordinate) was then binned 3 x 3 on the spatial axis and evaluated along the wavelength axis by a non-linear least squares fit of the temperature-dependence of simulated LIF spectra. For 60 bar, relative temperature distributions are inferred from a two-line NO-LIF measurement which was then calibrated to the measured temperature at 40 bar.

3. Results

The results of the imaging measurements are presented in figure 1 and in table 1 for various pressures in the $\phi = 0.95$ flame. Time-averaged band-pass filtered raw signals obtained with NO-excitation at 226.03 nm are shown in row A of figure 1. Only the central cone of hot exhaust gases above the flat flame front is stable; the outer part is increasingly fluctuating at high pressures and mixing with coflowing air. Because not all the corrections are linear in their temperature dependence, significant systematic errors affect the signal interpretation in the fluctuating regions. We therefore constrain the quantitative investigation of the various effects to the stable region visible in row B as a darker triangle based on a statistical analysis of signal standard deviation.

3.1. Fluorescence background

Row C of figure 1 shows the off-resonant signal (227.48 nm excitation) due to O_2-LIF and diffusely scattered light from the laser beam. A broadband LIF signal of unknown origin recently identified in the same burner [12] might also contribute to the background. Despite the fact that the best possible transition was chosen to discriminate against O_2-LIF [10], a contribution of background to the over-all signal of up to 29% at 60 bar is present in the center of this slightly lean flame (cf. table 1).

3.2. Temperature and quenching influence

Temperature influences the NO-LIF signal due to variations in ground state population (described by the Boltzmann function), line broadening and –shifting, and fluorescence quenching (through temperature dependence of collisional frequencies, quenching cross sections, and gas composition). The measured temperature fields are used to calculate the corrections for these effects based on a spectral simulation code [9] that includes line-broadening and shifting models [13-16] and quenching models [17].

Row D displays temperature fields for the respective flames obtained from the multi-line fitting procedure described in the experimental section. Measured temperatures are also given in table 1. The overall dependence of the NO concentration measurement on local temperature variations is shown in rows E and F of figure 1. During data evaluation, the temperature is reduced by 200 K compared to measured temperatures and the NO quantification is repeated. The images show the variation $1 - [c_{NO}(T - 200 \text{ K}) / c_{NO}(T)]$, and the numbers for a position 3 mm above the burner matrix are given in table 1.

For the NO A-X(0,0) excitation feature used in this work, the temperature sensitivity is minimized when calculating number densities (row E) since here, the temperature dependence of quenching and the overall temperature dependences of the ground state population and line broadening mostly cancel [9]. The T-sensitivity increases when calculating NO mole fractions (row F) due to the additional $1/T$ dependence [10]. With increasing pressure the measurements get more robust due to the loss of fine structure of the absorption spectra. The results show that in the central region a T-reduction by 200 K (for example in a turbu-

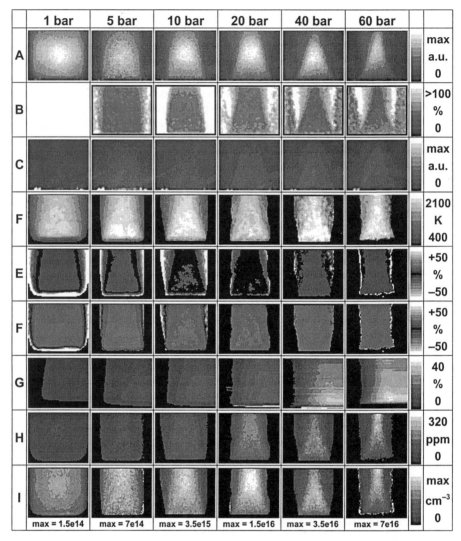

Fig. 1: Results of the imaging analysis from the methane/air flame ($\phi = 0.95$, without NO seeding) at various pressures: A. raw NO-LIF images (not corrected for laser sheet inhomogeneities), B. standard deviation derived from 250 instantaneous images (no data available at 1 bar), C. raw off-resonant images (not corrected for laser sheet inhomogeneities, same color code as in A), D. absolute temperatures from multi-line NO-LIF rotational thermometry measurements (for 60 bar: see text), sensitivity of the NO number density (E) and mole fraction measurements (F) on temperature variations of −200 K, G. total attenuation due to laser and signal absorption by CO_2 and H_2O, H. quantitative NO mole fractions, I. quantitative NO number densities. Further details are discussed in [9].

lent flow) would change the measurement of NO number density by $\leq 6\%$ at all pressures. However, when calculating NO mole fractions, the error with 200 K T-uncertainty increases to a nearly constant 14% from the additional $1/T$ dependence. The overall temperature sensitivity between $1200 - 2500$ K is $\pm 7 - 18\%$ (dependent on pressure) for measurements of NO number densities and becomes $\pm 38 - 50\%$ for measurements of NO mole fractions [9].

3.3. Signal and laser absorption

In high-pressure high-temperature combustion environments the UV transmission is reduced by absorption by majority species like CO_2 and H_2O [18]. This attenuates both the laser and emitted fluorescence intensities. Absorption cross-sections strongly depend on temperature and wavelength [7,19]. The distributions of H_2O and CO_2 are well known in the burnt gas of a laminar flame with cylindrical geometry. Therefore, in this flame with equilibrium burnt gas composition, the resulting correction factors depend only on the temperature distribution. Absorption by hot oxygen and self-absorption by NO are negligible [9].

Row G displays the calculated variation in transmission including the effects of laser attenuation (at 226 nm) and signal attenuation (at 237 nm) based on CO_2 and H_2O absorption cross-sections, using the measured temperature fields and the cylindrical symmetry of the flame as input. Despite the small size of the flame, total attenuation up to 43% is observed at 60 bar. The results for all flames at 3 mm above the burner matrix are given in table one.

3.4. Calibration

The calibration information is obtained from NO-LIF measurements in flames with NO seeding between 300 and 600 ppm [4,5,9,20]. The NO-LIF signal is linearly proportional to NO seeding in slightly lean flames [8] and the in-situ calibration measurement minimizes uncertainties due to variations in pressure, temperature, and gas composition. In our experiment, NO-addition measurements are carried out for each pressure independently.

Figure 2 shows the calibration curves at different pressures. The measured signal intensities are corrected for attenuation and pressure and temperature dependencies as described above and plotted versus seeding NO number densities. The calibration information which links measured pixel counts with local NO number densities is obtained from the slopes only. Varying offsets indicate different natural NO concentrations at the measurement location for the various flame conditions. The measurement position (3 mm above the burner matrix) is below the point of maximum NO concentration at 60 bar, which reduces the offset, but close to the maximum in the 40 bar flames, resulting in an increased offset. The second plot shows the slopes of the calibration curves with pressure. Because they are corrected for all T- and p-effects, all the slopes should be identical. The variation ($\pm 10\%$) indicates the uncertainty from calibration at different pressures (errors in the spectra simulation, variation

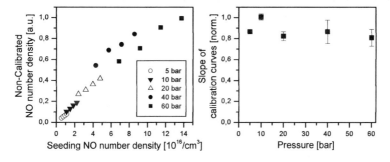

Fig. 2: Calibration by NO seeding: The left plot shows the linearity of the NO signal intensity versus NO seeding concentration at 5, 10, 20, 40 and 60 bar. The slopes of the resulting linear fits yield the calibration information and indicate the over-all NO-LIF yield. The right plot compares these data with the results of spectra simulations taking into account T and p-dependent effects.

Table 1: Results of the imaging analysis in the central region 3 mm above the burner matrix. The temperature sensitivity is the variation in evaluated mole fraction or number density when lowering the measured temperature by 200 K (see text).

Pressure [bar]	Off-resonant signal [%]	Measured temperature [K]	T-sensitivity mole fraction [%]	T-sensitivity number density [%]	Absorption correction [%]	Measured NO mole fraction [ppm]	Measured NO number density [1/cm^3]
1	5	1870 ±6%	−16	−6	0.7	23.8 ±11%	9.22 x 10^{13} ±11%
5	15	1981 ±6%	−14	−4	4	27.7 ±11%	5.07 x 10^{14} ±11%
10	9	1968 ±7%	−12	−2	7	66.4 ±11%	2.45 x 10^{15} ±11%
20	11	1862 ±8%	−12	−1.5	11	137.2 ±11%	1.06 x 10^{16} ±11%
40	22	1967 ±9%	−14	−4	24	180.2 ±12%	2.62 x 10^{16} ±11%
60	29	1942 ±15%	−15	−5	29	141.8 ±13%	3.16 x 10^{16} ±11%

in NO reburn, and overall stability of the experiment), thus assessing the error introduced by calibration at a single fixed pressure for an experiment such as an IC engine with varying pressures [9].

3.5. Quantitative NO concentrations

The last two rows of fig. 1 show the quantitative images of natural NO in the high-pressure flame as mole fractions (row H) and number densities (row I). These images are corrected for laser and signal attenuation and for all p and T-dependent terms discussed above. Then they are calibrated using the measured calibration shown in figure 2. The resulting natural NO concentrations are listed in table 1.

The over-all accuracy for the quantitative measurements is ±11% for the NO number density measurements. The main contribution to the error is the systematic uncertainty of the calibration due to reburn of the seeded NO and overall experimental stability (estimated: ±10%, cf. figure 2). Errors in the measured temperature (resulting uncertainty on NO number density: ±1 – 2%), the LIF measurement itself (ca. ±3%) and the spectral simulation code used for correcting the data (ca. ±2%), add up to 4 – 5% (dependent on pressure). The over-all accuracy decreases only slightly when determining NO mole fractions instead of number densities.

4. Conclusions

The steady laminar flame allows corrections for non-resonant background contributions, temperature, and signal and light absorption. The impact of each correction is quantified to assess the errors introduced in NO measurements in practical high-pressure combustors, where simultaneous measurements of temperature, background fluorescence and laser and signal attenuation is not feasible.

Under the conditions of the slightly lean flame, the background contribution which remains after tuning the laser off the NO resonance is nearly 25% (at 60 bar, relative to a natural NO concentration of 165 ppm) and is a significant contributor to the over-all error when no corrections can be made. The total sensitivity of the NO number density measurements on temperature variations is surprisingly low because the effects of quenching (including the temperature-dependent variation of equilibrium burnt gas compositions) and the T-dependence of the spectra almost cancel. However, mole fraction determination requires local temperature, and the resulting uncertainty when no information on local temperatures is available increases significantly. In applications with variable pressure (e.g. internal combustion engines) calibration is typically made at a single, fixed pressure. As long as the pressure-dependent effects are included in the data evaluation (rather then calibrating at each individual pressure) measurement uncertainties are not significantly increased.

Absorption by hot combustion products (CO_2 and H_2O) has often been ignored in previous work. Despite the small size of our flame (8 mm diameter), the total attenuation of laser excitation and signal light at 60 bar is up to nearly 45%. Thus, absorption is of major concern for quantitative NO measurement, especially in larger scale practical applications.

Acknowledgements

Work at Stanford supported by the US Air Force Office of Scientific Research, Aerospace Sciences Directorate, with Julian Tishkoff as the technical monitor. The Division of International Programs at the US National Science Foundation supports the Stanford collaboration via a cooperative research grant. The University of Heidelberg work and the travel of WB and CS are sponsored by the Deutsche Forschungsgemeinschaft (DFG) and the Deutsche Akademische Auslandsdienst (DAAD).

References

[1]   Wolfrum J 1998 *Proc. Combust. Inst.* **27** 1
[2]   Reisel J R and Laurendeau N M 1994 *Combust. Sci. and Tech.* **98** 137
[3]   Ravikrishna R V and Laurendeau N M 2000 *Combust. Flame* **120** 372
[4]   Josefsson G, Magnusson I, Hildenbrand F, Schulz C and Sick V 1998 *Proc. Combust. Inst.* **27** 2085
[5]   Bessler W G, Schulz C, Hartmann M and Schenk M 2001 *SAE Technical Paper Series* 2001-01-1978
[6]   DiRosa M D, Klavuhn K G and Hanson R K 1996 *Comb. Sci. Tech.* **118** 257
[7]   Schulz C, Koch J D, Davidson D F, Jeffries J B and Hanson R K 2002 *Chem. Phys. Lett.* **355** 82
[8]   Schulz C, Sick V, Meier U, Heinze J and Stricker W 1999 *Appl. Opt.* **38** 1434
[9]   Bessler W G, Schulz C, Lee T, Shin D.-I., Hofmann M, Jeffries J B, Wolfrum J and Hanson R K 2002 *Applied Physics B* (in press)
[10]   Bessler W G, Schulz C, Lee T, Shin D.-I., Jeffries J B and Hanson R K 2002 *Appl. Opt.* (in press)
[11]   Vyrodow A O, Heinze J, Dillmann M, Meier U E and Stricker W 1995 *Appl. Phys. B* **61** 409
[12]   Bessler W G, Schulz C, Lee T, Jeffries J B and Hanson R K 2002 *Appl. Opt.* (submitted)
[13]   Chang A Y, DiRosa M D and Hanson R K 1992 *J. Quant. Spectrosc. Radiat. Transfer* **47** 375
[14]   DiRosa M D and Hanson R K 1994 *J. Mol. Spectrosc.* **164** 97
[15]   DiRosa M D and Hanson R K 1994 *J. Quant. Spectrosc. Radiat. Transfer* **52** 515
[16]   Vyrodov A O, Heinze J and Meier U E 1995 *J. Quant. Spectrosc. Radiat. Transfer* **53** 277
[17]   Paul P H, Gray J A, Durant Jr J. L. and Thoman Jr. J W 1993 *Appl. Phys. B* **57** 249
[18]   Hildenbrand F and Schulz C 2001 *Appl. Phys. B* **73** 173
[19]   Schulz C, Jeffries J B, Davidson D F, Wolfrum J and Hanson R K 2002 *Proc. Combust. Inst.* **29** (in press)
[20]   Mokhov A V, Levinsky H B and van der Meij C E 1997 *Appl. Opt.* **36** 3233-3243

Inst. Phys. Conf. Ser. No. 177
Paper presented at 1st Int. Conf. on Optical & Laser Diagnostics, London, 16–20 Dec. 2002
©2003 IOP Publishing Ltd

Spray characteristics of a multi-hole injector for direct injection gasoline engines

N Mitroglou, J M Nouri and C Arcoumanis,

Centre for Energy and the Environment, School of Engineering and Mathematical Sciences, City University, London

Abstract. The sprays from a high-pressure multi-hole injector into a constant volume chamber have been visualised and quantified in terms of droplet velocity/diameter temporal and spatial distributions at injection pressures up to 200 bars and chamber pressures varying from atmospheric to 12 bar. The overall jet spray angle relative to the axis of the injector was found to be 40o and almost independent of injection and chamber pressure. A significant advantage relative to swirl pressure atomisers.

The temporal profile 10 mm from injector showed that droplet velocities increased sharply at the start of injection to maximum values 120 m/s and then remained unchanged during the main part of injection before decreasing rapidly towards the end of injection. The spatial velocity profiles were jet-like at all axial locations with the maximum at centre of the jet. The Sauter mean diameters in the main spray 10 mm from injector were of order of 19 and 14 µm at injection pressure of 120 and 200 bars and under atmospheric condition. Within measured range, the effect of injection pressure on droplet size was small while the increase in chamber pressure resulted in much smaller droplet velocities by up to fourfold and larger droplet sizes by up to 40%.

Inst. Phys. Conf. Ser. No. 177
Paper presented at 1st Int. Conf. on Optical & Laser Diagnostics, London, 16–20 Dec. 2002
©2003 IOP Publishing Ltd

Hybrid optical/digital codec for mitigation of optical aberrations

Samir Mezouari and Andrew R Harvey

Computing and Electrical Engineering department, Heriot-Watt University, Edinburgh, EH14 4AS, Scotland, UK.

Abstract. To extend the depth-of-focus of incoherent imaging systems an aspherical pupil plane element is employed to encode the incident wavefront in such a way that the image recorded by the detector can be accurately restored over a large range of defocus. This approach alleviates the defocus and related aberrations whilst maintaining a diffraction-limited resolution for incoherent imaging systems. This offers the potential to implement diffraction-limited imaging systems using simple and low-cost one- or two-element achromatic athermal lenses. However these performance are associated with reductions in signal-to-noise ratio of the displayed image. This effect sets a serious limit to future applications, particularly in thermal imaging where the optical systems involve fast optics. Fast optics means a small depth of field andsignal-to-noise ratio of the deconvolved image can be reduced significantly for small amounts of defocus. Fortunately, recent and future improvements in detector sensitivity offer some scope for allowing modest amounts of noise amplification whilst maintaining current detection performance levels. The system performance of the present hybrid optical/digital technique, called wavefront coding, depends substantially on the proper design of the phase mask. Merit functions are derived to explore the optimum design.

1. Introduction

The main task of any optical designer is to produce an image located in a single plane (the nominal detector plane) by focusing light with the use of optical elements. Unfortunately, images formed at a single image-plane suffer from aberrations due to imperfections in lens action. The deviation from an ideal diffraction-limited image is associated with familiar aberrations, such as thermal and chromatic defocus, coma, and astigmatism. For industrial and military use, imagers are designed to work in hostile environments where the temperature range can be large. This is a particular problem in thermal imaging where the optical components are commonly made with semiconductor materials, such as Germanium, which have a large temperature coefficient of refractive index. Therefore, an optical system with a large tolerance to defocus aberration will suffer less from chromatic and thermal aberration. The conventional approach to alleviate the defocus aberration is either by introducing additional and complex optical elements, or by including movable corrective mechanisms. Although the original idea of using an hybrid optical/digital system to extend the depth of focus of incoherent imaging systems has been first suggested by Haustler [1] in 1972, a significant success has been achieved only recently by Dowski et al [2], who employed an aspherical phase mask to encode the transmitted wavefront in such a way that the point spread function (PSF) is invariant close to the image plane. In this paper we explore the performance of this modern technique, called *Wavefront Coding*, with a particular emphasize of the noise amplification associated with application of this technique.

2. Background

According to Hopkins [3], the primary defocus aberration can be described by the parameter W_{20} and in the simple case of a square aperture, it is given by [4]

$$W_{20} = \frac{a^2}{8}(\frac{1}{z} - \frac{1}{f}), \tag{1}$$

where a represents the aperture width f is the focal length and the distance between the aperture plane and the image plane is given by z. When $z=f$, the image plane is in-focus and $W_{20}=0$ (system well focused). Rayleigh demonstrated that the tolerance to defocus aberration is achieved when $W_{20}<\lambda/4$ (Rayleigh's quarter-wave rule). This limit on the degree of defocus corresponds to an image formed with quality that is only slightly inferior to that with aberration-free system ($W_{20}=0$). The phase mask that achieves a large depth of field has been derived from the evaluation of the optical transfer function (OTF) of the system by using a stationary phase approximation. The rectangularly separable generalised pupil function with respect to the normalised spatial coordinates is given by

$$P(x,y) = \exp[j\alpha(x^3 + y^3)], \quad \text{for } |x|\leq1, |y|\leq1 \tag{2}$$

where the parameter α controls the peak-to-valley optical path difference introduced by the pupil-phase mask. This *cubic* phase mask is anti-symmetric and rectangularly separable. The tolerance to defocus aberration is shown in figure 1 where the computed modulation transfer functions (MTF) are displayed for various values of W_{20} and compared with the standard optical system ($\alpha=0$). The values of W_{20} are normalised with respect to a wavelength. The tolerance to defocus increases with the increase of the values of α. For $\alpha>9\pi$, the MTF is insensitive to a defocus of up to $W_{20} =3$, and contains no zeros. However, the MTF is reduced for high spatial frequencies as α increases. The gain in tolerance to defocus is achieved at the expense of reducing the MTF and this is associated with a lower SNR in the decoded image. Therefore, the noise amplification sets a limitation to the performance for this technique. In the next section, we explore the noise amplification as a function of the extended depth of focus.

3. Noise amplification

The image recorded by the detector is encoded by convolution with the point spread function of the mask. The decoded image is obtained by applying a deconvolution technique to restore fully the information and produce a clear image. Assuming that the optical system is space-invariant, the recorded image, in the Fourier domain, is given by

$$I(f_x,f_y) = O(f_x,f_y)\, OTF(f_x,f_y) + N(f_x,f_y), \tag{3}$$

where O is the object spatial frequency distribution, and I is the Fourier transform of the intermediate image recorded by the detector. N is the noise in the system. The modelling presented in this paper is performed with white noise. The noise amplification showed in Figure 2 is obtained with simple inverse digital filtering. The calculation has been undertaken with a Fresnel number of 50. When $\alpha=0$ the optical system corresponds to a simple rectangular aperture and so there is no amplification of noise. As expected, the noise amplification increases as the phase-delay introduced by the phase mask increases and the modulation trasfer function is reduced. The variation in noise amplification with α is shown in in Figure 2: it is composed of an approximately linear component and an oscillation. suggesting that the noise amplification seems to follow a linear regression, so the noise amplification is approximately proportional to the peak-to-valley optical path difference, α, of the phase mask. Similarly, the extended depth of field accomplished by the phase mask is calculated by using a quality factor [5] that is equivalent to the Rayleigh's quarter-wave rule for aberration tolerance. Figure 3 shows a similar oscillation as in Figure 2 close to $\alpha=3\pi$. Therefore, the phase mask achieves nearly the similar performance within the interval [5/2 π, 7/2 π] and this response occurs regularly as α increases, however, we consider, within a good approximation, that the depth of focus vary linearly with the parameter α. It will be shown that the depth of field increases in proportion to α and thus, the enhancement in the depth of focus obtained by the wavefront coding is proportional to the noise amplification in the final image. To illustrate the performance of the wavefront coding technique, we discuss a simple example. The merit figures shown in this paper represent a good starting point for a designer to fit the parameters of the phase mask for particular specific requirements.

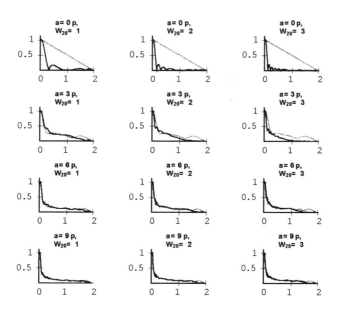

Figure 1. The variation of the computed MTF for different values of α and W_{20}

Traditionally, the tolerance to defocus is improved by reducing the aperture size. For instance, when a pupil aperture is reduced to a half of its initial size the tolerance to defocus is improved by a factor of 4 according to Rayleigh's rule (see equation 1). Since the cut-off frequency is proportional to the aperture width, it is reduced by a factor of 2. Hence, the standard approach to control the defocus aberration is governed by the resolution/extended depth of focus trade-off. According to the merit function displayed in Figure 3, an improvement of a factor 4 in the depth of focus is achieved by a cubic phase mask that has a parameter $\alpha= 2\pi$. When the cubic phase mask ($\alpha= 2\pi$) is added in front of the full pupil aperture, the restored image obtained with the wavefront coding system has a signal-to-noise ratio reduced by a factor of approximately 5 (see Figure 2), but without any loss in resolution.

In conclusion, the wavefront coding technique alleviates defocus errors with its related aberrations while maintaining the diffraction-limited performance according to the noise amplification/extended depth of focus trade-off.

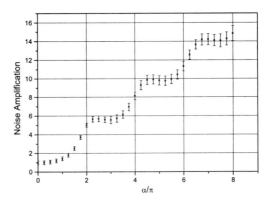

Figure 2. The noise amplification resulting from different optical phase difference α

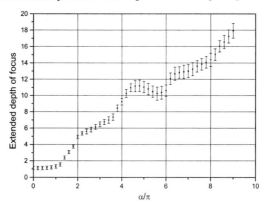

Figure 3. The extended depth of field as a function of α

This work was supported by QinetiQ, Malvern, United Kingdom.

References
[1] G. Häustler, "A method to increase the depth of focus by two steps image processing," Opt. Commun. 6, 38-42 (1972)
[2] E. R. Dowski, Jr. and W. T. Cathey, "Extended depth of field through wavefront coding," *Appl. Opt.* 34, 1859-1866 (1995).
[3] H. H. Hopkins, "The use of diffraction-based criteria of image quality in automatic optical design," *Optica Acta* 13, 343-369 (1966)
[4] Joseph W. Goodman, "Introduction to Fourier optics," McGraw-Hill international editions (2nd edition) pp 148, (1996)
[5] S. Mezouari, A. Harvey, "Wavefront coding for aberration compensation in thermal imaging systems," *SPIE annual Meeting*, San Diego, 3-5 August 2001.

Inst. Phys. Conf. Ser. No. 177
Paper presented at 1st Int. Conf. on Optical & Laser Diagnostics, London, 16–20 Dec. 2002
©2003 IOP Publishing Ltd

Effect of chamber pressure on the spray structure from a swirl pressure atomiser for direct injection gasoline engines

J M Nouri* and J H Whitelaw**

* School of Engineering and Mathematical Sciences, City University, London EC1V 0HB
**Mech. Eng. Dept., Imperial College of Science, Technology and Medicine, London SW7 2BX

Abstract. Images and droplet velocities/diameters of the spray generated by a swirl pressure atomiser have been measured inside a chamber with back pressures ranging from atmospheric to 12 bar and injection pressures up to 100 bar. The photographic investigation showed that the spray structure consists of a non-swirling pre-spray and the main spray whose structure was found to be very sensitive to chamber pressure. In general, the spray cone was contracted and had smaller tip velocity and, thus, reduced penetration at higher chamber pressures due to the enhanced drag; an increase in chamber pressure from 1 to 12 bar caused an overall reduction in the spray cone angle by 35% and in tip velocity by 50%. Measurements by phase-Doppler anemometry (PDA) confirmed the observed in the images effect of chamber pressure (up to 12 bar) and showed a similar reduction in droplet mean velocities to that obtained from flow visualisation, i.e. a suppression in droplet rms velocities by up to 40%, an increase in droplet Sauter mean diameter by up to 20%.

1. Introduction

Charge stratification for overall lean-burn combustion in direct-injection gasoline engines is achieved by injecting fuel during the compression stroke, and increased volumetric efficiency by injection on the intake stroke, Iwamoto et al. (1997) and Wirth et al. (1998). In both cases the fuel spray characteristics plays a major role and has to be suitable for full load and stratified charge in particular the latter case where the formation of a stable and ignitable fuel mixture required around the spark plug. Thus, knowledge of spray characteristics, including spray shape, penetration and droplets velocities/diameters is required as functions of time and injection and chamber pressures.

Previous investigations under high back pressure are limited, for and include those of Shelby et al. (1998), Le Coz (1998), Wigley et al (1998) and Gavaises et al (2002). The first two studies showed that spray angles decreased with chamber pressure and the second two presented local measurements of the droplet velocities with limitations imposed on the optical methods by high droplet concentrations. The visualisations of Nouri et al (1999) and Shelby et al (1998) showed that the spray shape was independent of injection pressure and Ipp et al (1999), with two-dimensional Mie and LIF techniques, and showed that the cone angle and tip penetration reduced with increase in back pressure. It is also known from the experiments of Xu and Markle (1998) that air drag increased with chamber pressure and decelerated fuel droplets with increased probability of coalescence and the formation of larger droplets. Nouri and Whitelaw (2001) quantified the effect of injection duration on a pressure swirl injector and showed poor atomisation for injection duration of 0.15 ms or less. Ortmann et al (2001) examined two different types of injectors for gasoline direct injection engines and concluded multi-hole injectors are more stable than swirl type.

The present investigation of the spray generated by a swirl injector for gasoline engines extends that of Nouri et al. (1999) and Nouri and Whitelaw (2001) in that it quantifies the effect of increasing chamber pressure from atmospheric to 12 bar. The spray was visualised through its central plane with a laser sheet and a CCD camera, and the droplet velocities and size were measured with a phase-Doppler anemometer. The following sections describe, the experimental arrangement and instrumentation, the results and their implications and summary conclusions.

2. Experimental arrangement and instrumentation

The injector was a prototype (A) provided by FIAT/CRF with a single exit nozzle. Swirl was created inside the discharge hole by two off-centre tangential slots, which guided the fuel from the nozzle gallery. The fuel (2,2,4 Trimethylpentane with a density, kinematic viscosity and surface tension of 692 kg/m³, 0.78 cSt and 0.0188 N/m, respectively) was supplied from three pressurised fuel vessels connected to each other by a high-pressure rail. The vessels were filled with the fuel and pressurised from a nitrogen cylinder with a polymer diaphragm separating gas and liquid fuel as shown in figure 1. The system was designed to supply fuel to the injector with pressures up to 100 bars and a regulator connected to the cylinder ensured no visible change in pressure from the gauge. The injector sprayed vertically downwards into a pressure chamber with injection pressure up to 100 bar, chamber pressures of atmospheric, 6 and 12 bars and an injection duration of 3.2 ms. The injection frequencies were 1 and 0.2 Hz for the spray imaging and phase Doppler measurements, respectively.

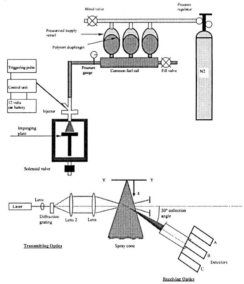

Figure 1 The high-pressure fuel system and optical configuration of the phase-Doppler anemometer.

The inside dimensions of the pressure chamber were 80x80x140 mm with volume similar to that of typical passenger cars. The chamber had two windows on the opposite side of the chamber with viewing dimensions of 80x110 mm and a third window of dimensions 10x110 mm on the third side of the chamber. The windows were made of Perspex with a refractive index of 1.49. In order to avoid spray fouling, the chamber was vented after each injection by a solenoid valve at the exit of the chamber with an injection frequency of 0.2 Hz, i.e. 5 s. The solenoid valve was activated by a delayed time pulse (around 0.5 second after the start of the injection) and remained on for 1s. The rest of the time (3.5 s) was to allow the built up of the pressure inside the chamber before the next injection.

Instantaneous images of spray with a time resolution of 5 μs, were obtained by a CCD camera equipped with an intensifier and a 50 mm Nikon-lens, and included overall images and close-ups with higher magnification achieved with Nikon adapters of 13 and 30 mm. The CCD camera was synchronised with the injection pulse so that the images corresponded to particular times after the start of the needle lift, hereafter referred to as start of the injection. This time was different from that of the emergence of the initial spray from the exit hole of the injector due to the injector design and voltage applied to the coil, plus the time taken for the fuel to pass through the nozzle passage to its exit; this delay time was obtained from images to be around 0.225 ms. A plane light sheet was produced by an argon-ion laser (Spectra Physics) in the green mode at a constant power of 600 mW and a cylindrical lens with 6 mm focal length; the width of the laser sheet was around 1 mm and was arranged to pass through the centre of the spray.

The phase-Doppler anemometer was similar to that described by Nouri and Whitelaw (2001) with light collected at 30 degrees to the axis of the transmitted beams. The argon-ion laser was tuned to the green wavelength at 514.5 nm with power of around 400 mW, a rotating diffraction grating divided the incident beam into two equal intensity beams and provided frequency shift. The intersection volume had major and minor axes of approximately 1.97 and 0.084 mm. Light scattered by the droplets was collected by a 310 mm focal length positioned at 30 degrees to the plane of the incident beams to ensure that refraction dominated the scattered light. The light was then focused to the centre of a slit with a width of 0.1 mm to provide an effective length of the measuring volume of 0.387 mm. The scattered light was passed through a mask with three evenly spaced rectangular apertures to three photodetectors (Hamamatsu Model R-1477).

Velocity, size and time, information was collected continuously over many injection cycles from which ensemble-averages were obtained over a time window of 0.2 ms was found to be sufficient. The total number of samples ranged from 25,000 to 30,000, and the number of validated samples in the 0.2 ms time interval varied from 200 to 400 samples with maximum statistical uncertainties of around 2% in the ensembled mean and 10% in the rms of the velocity fluctuations, based on 95% confidence and 20% turbulence intensity, Yanta (1973). The maximum frequency ambiguity for the present counter with its 20 MHz clock was 63 kHz or 0.38 m/s.

Detailed accounts of the uncertainties and limitation associated with the PDA measurements are given by, for example, Wigley (1993) and Hardalupas et al (1994). An important source of uncertainty in the near-injector region was the attenuation of the laser beams and the scattered light due to high concentrations of droplets. The extent of the turbidity of the spray with the present injector was evident up to 60 mm from the injector as shown by Nouri et al. (1999). The phase-Doppler results presume spherical droplets and, since sprays are known to include ligaments in the near field, this is a potential source of uncertainty very close to the injector. The photographic investigation suggested that there were very few, if any, ligaments in the regions of present measurements and the verification system of the counter should have rejected non-spherical droplets.

3. Results and discussion

The flow-visualisation photographs and measurements of the droplet velocity and size are presented and discussed in the following two subsections. The volumetric capacity of the injector was determined as a function of injection pressure by measuring the volume of the fuel from up to

5000 injections into a graduated container with a resolution of 0.5 ml, with repeatability better than 2%. Figure 2 presents the variation of the fuel volumetric capacity of the injector as a function of injection pressure under atmospheric condition and for injection duration of 3.2 ms. The result shows that the injected volume per pulse varies almost linearly with increase in injection pressure and that at an injection pressure of 80 bar the volumetric capacity of the injector was found to be 0.0575 cc per injection; this is equivalent to a volume flow rate of 18 cc/sec per injection The volume flow rate is expected to be less with higher chamber pressures so that a reduction of order of 7.3 is expected with injection pressures of 80 bar when the chamber pressure increased from atmospheric to 12 bar.

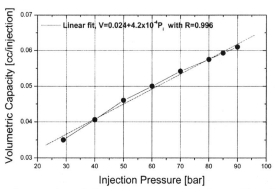

Figure 2 Volumetric capacity of the injector as a function of injection pressure under atmospheric conditions and at injection duration of 3.2 ms.

3.1 Flow visualisation

Figure 3 shows the overall images of spray development as a function of time after the start of injection at three chamber pressures of 1, 6 and 12 bars. Under atmospheric condition, figure 3(a), the initial angle of the cone spray was found to be about 62 degrees; this was determined from close-up images and within the first 10 mm from the injector at a time in the middle of injection when the spray was stable. The images show two distinct phases of spray development, first the initial spray, which emerges in the centre as a long column containing fuel ligaments and droplets. This is mainly because the initial fuel does not have enough tangential velocity to form a cone when the needle is first lifted and, therefore, emerges along the axis of the injector. The main spray emerges after the initial phase as a dense cone spray. It is evident that the number density of droplets close to the edge of the spray is greater than in the central region due to higher intensity of the scattered light; it seems that as the spray convected downstream, the droplets become more uniformly distributed across the cone spray. The contraction of the spray away from the injector is due to the low-pressure region created in the centre of the spray and the air entrainment into the spray from surrounding. A vortical-shape structure around the periphery at the leading edge of the main spray is also evident.

Figure 3(b) and (c) shows images of spray at the same times of injection and the same injection pressure as those of figure 3(a), but at chamber pressures of 6 and 12 bars. Like under atmospheric condition, the results show two phases of spray development although the initial phase seems to be suppressed as the chamber pressure is increased. The effect of increasing chamber pressure is markedly evident with large reduction is spray width and spray penetration length. The images show clearly that the spray penetration and width get shorter and narrower with increase in chamber pressure. For example the width of spray 10 mm from the injector is 10.5 mm at atmospheric pressure, which reduces to 6.5 mm at chamber pressure of 12 bar. This suggest an overall reduction of about 38% that is equivalent to a reduction of 62% in cross-section area or 35% in spray cone angle, which implies a denser spray and probably more uniformly distributed droplets number density. Also at 3 ms after the start of injection the spray tip penetration is about 102 mm under atmospheric condition, which reduces to 69 mm when the chamber pressure is increased to 12 bar, i.e. 32% reduction. The average tip velocity was determined from measurements of the distance travelled by the tip of the spray from 0.6 to 1.0 ms after the start of injection and found to be 69 and 35 m/s for chamber pressures of atmospheric and 12 bar, respectively, i.e. a reduction of about 50% due to increased air drag.

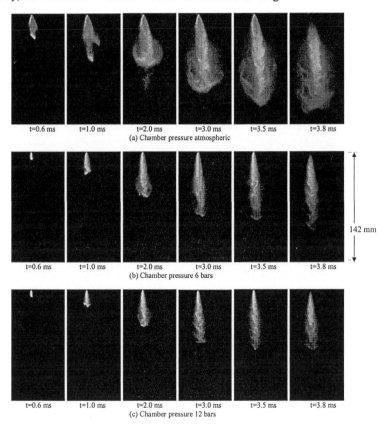

t=0.6 ms   t=1.0 ms   t=2.0 ms   t=3.0 ms   t=3.5 ms   t=3.8 ms
(a) Chamber pressure atmospheric

t=0.6 ms   t=1.0 ms   t=2.0 ms   t=3.0 ms   t=3.5 ms   t=3.8 ms
(b) Chamber pressure 6 bars

142 mm

t=0.6 ms   t=1.0 ms   t=2.0 ms   t=3.0 ms   t=3.5 ms   t=3.8 ms
(c) Chamber pressure 12 bars

Figure 3 Images of spray structure as a function of time after the start of injection at an injection pressure of 80 bar and an injection duration of 3.2 ms for different chamber pressures of (a) atmospheric, (b) 6 bar and (c) 12 bar.

A sequence of close-up images 2 ms after the start of injection as a function of injection pressure is presented in figure 4 with magnification more than 13 times that of figures 3. At the lower chamber pressure, figure 4(a), and at an injection pressure of 3 bar, the spray has formed a hollow cone with a cluster of ligaments and droplets in the annulus. The cluster disintegrates into droplets as the injection pressure increases, and fully atomised sprays are formed at injection pressures above 30 bar; similar spray structure was observed by Shelby et al. (1998). The images also show that some fuel droplets tended to move radially inwards, possibly due to low tangential velocity and low-pressure region in the centre. It is also evident that the cone angle is almost independent of the injection pressure as it depends only on the ratio of axial to tangential velocities and the geometry of the discharge hole. At the higher chamber pressure of 12 bar and injection pressures of 13 and 14 bars, figure 4(b), the spray does not show a hollow cone structure with most of the fuel in the form of liquid sheet or ligaments. As the injection pressure increases more of the fuel is atomised so that at an injection pressure of 30 bar the spray seems to be fully atomised.

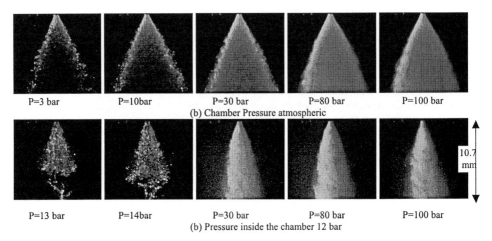

P=3 bar   P=10bar   P=30 bar   P=80 bar   P=100 bar
(b) Chamber Pressure atmospheric

P=13 bar   P=14bar   P=30 bar   P=80 bar   P=100 bar
(b) Pressure inside the chamber 12 bar

10.7 mm

Figure 4 Images of spray 2ms after the start of the injection at different injection pressures, injection duration of 3.2 ms, and two chamber pressures (a) atmospheric and (b) 12 bar.

3.2 Droplet velocity and size distribution

The droplet velocities and size were obtained by the PDA system at different axial distances from the injector for two chamber pressures of atmospheric and 12 bar at an injection pressure of 80 bar and an injection duration of 3.2 ms. The ensemble average values of mean and rms velocities and the AMD and SMD droplet sizes were resolved over 0.2 ms time intervals. A sample of results is presented to quantify the effect chamber pressure.

Figure 5 presents temporal variation of droplets velocity and diameter in the annulus of the cone spray at an axial location 80 mm from injector. In general, for both chamber pressures, the temporal variation of droplets velocities shows clearly the transient nature of the spray with two distinct velocity variations so that the droplet velocity decreases to a minimum with time in the initial phase. Then the second phase starts with the arrival of the main spray and the droplet velocity increases with time to a maximum before it reduces to almost zero towards the end of injection, with large velocity fluctuations as previously observed by Nouri et al. (1999), Wigley et al. (1998) and Shrimpton et al. (1997). The size distribution showed a gradual decrease in droplets sizes from the leading to the trailing edge of the spray and almost constant values in the tail of the spray. The AMD and SMD values were around 18 and 25 μm during the main part of the spray under atmospheric condition. These values were increased to 22 and 30 μm at chamber pressure of 12 bar.

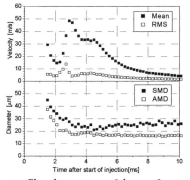 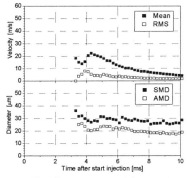

(a) Chamber pressure 1 bar, r=6 mm. (b) Chamber pressure 12 bar, r=4 mm

Figure 5 Comparison of temporal profiles of droplets velocity and diameter distribution between two chamber pressures of atmospheric and 12 bar at an axial distance of 80 mm, an injection pressure of 80 bar and injection duration of 3.2 ms.

The effect of chamber pressure is clearly evident so that droplets mean velocity at the higher chamber pressure is much lower. For example the maximum droplet velocity at the chamber pressures of 1 and 12 bars are 48 and 23 m/s, respectively, i.e. a differences of 52% similar to that observed in spray images. The effect of chamber pressure on droplets size was small in the leading edge of the spray, but during the main part of spray the increase in chamber pressure from 1 to 12 bars increased the droplets diameters (both AMD and SMD) by up to 18%. The increase in diameters is consistent with the contraction of the spray at the higher chamber pressure and the increased probability of coalescence as was noted by Arndt et al (2001).

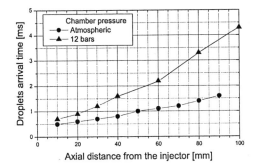

Figure 6 Variation of droplets arrival time on the spray axis as a function of the distance from the nozzle exit for chamber pressures of 1 and 12 bars at an injection pressure of 80.

A delay in droplets arrival time at this location at the higher chamber pressure is also evident, around 1.8 ms, which indicates a reduction in spray penetration; again similar to spray visualization. The delay in droplets arrival time with chamber pressure was further examined by

128

plotting the droplets arrival time on the axis of spray at different axial locations and the results are presented in figure 6. The results show that the delay in droplets arrival time increases with axial distance so that the delay at 20 mm from injector is 0.3 ms, which increases to 1.9 ms at 80 mm.

The radial distribution of droplet velocity and diameter at z=30 and 80 mm from injector in the trailing edge of the spray are shown in figure 7 for chamber pressures of 1 and 12 bars. The mean velocity profiles across the spray, at both locations and chamber pressures, exhibited a jet like distribution with the peak around the centre, while the rms velocity distribution was uniform with values of order of 3.5 and 2 m/s at z=30 mm and for chamber pressures of 1 and 12 bars, respectively; the corresponding values at z=80 mm are 5 and 3 m/s.

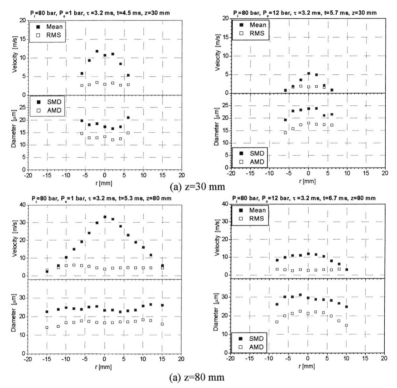

(a) z=30 mm

(a) z=80 mm

Figure 7 Comparison of radial profiles of droplet velocity and diameter distributions between two chamber pressures of 1 and 12 bar at a time corresponding to the trailing edge of the spray, for an injection pressure of 80 bar and injection duration of 3.2 ms; (a) z=30 mm, (b) z=80 mm.

The diameter profiles were also uniform across the section in the leading edge with SMD values of around of 18 and 22 μm at z=30 mm and for chamber pressures of 1 and 12 bars, respectively. Further downstream at z=80 mm, these values were become larger with corresponding values of 24 and 30 μm which may be caused by droplets collision. As the droplets are convected downstream they tend to migrate from annulus towards the low-pressure region at the centre, which enhance the possibility of droplets collision; also the inherent different trajectories of droplets may cause coalescence.

Figure 7 also show the effect of chamber pressure on droplets velocity and diameter across the spray with considerably reduction in mean velocity as the chamber pressure increased from 1 to 12 bars at both axial locations. As a result, the droplets velocity fluctuations have been suppressed at higher chamber pressure by about 40%. The diameter profiles also show an increase in droplets diameter with chamber pressure particularly around the central region with an average increase in SMD value of 18% when the chamber pressure was increased from 1 to 12 bars at z=30 mm; this increase at z=80 mm was about 20%.

4. Conclusions

The effect of back pressure on the spray structure generated by a swirl atomiser was quantified through spray visualisation and PDA measurements. The photographic investigation showed the transient nature of the sprays and that the shape of the sprays was nearly independent of injection pressure at all chamber pressures. The fully formed sprays comprised annular cones with an angle around of 60 degrees. The spray structure was found to be very sensitive to chamber pressure so that the spray cone was contracted and had smaller velocity and therefore less penetration with an increase in chamber pressure from 1 to 12 bars with an overall reduction in spray cone angle of 35% and tip velocity reduction of 50% due to higher drag.

Quantitative measurements with a phase-Doppler anemometer confirmed the photographic observations and showed the details of the non-swirling pre spray and the main spray. The effect of chamber pressure was considerable with large reduction in droplet mean velocity by up to 50%, similar to that observed from the images, large suppression of the droplet velocity fluctuations by 40% and an increase in the droplet arithmetic and Sauter mean diameters by up to 20 % when the ambient pressure inside the chamber was increased from 1 to 12 bars. The results also showed a delay in the droplets arrival time at the higher chamber pressure, which increased with axial distance from the injector.

ACKNOWLEDGEMENT
The financial support from the European Union under contract Joule JOF3-CT97-0028 is gratefully acknowledged.

References
Arndt S, Gartung K and Bruggemann D **2001** *Proc. ILASS-Europe 2001.*
Gavaises M, Abo-Serie E and Arcoumanis C **2002** *SAE* 2002-01-1136.
Hardalupas Y, Taylor A M K P and Whitelaw J H **1994** *Int. J. Multiphase Flow*, **20,** 233-259.
Iwamoto Y, Noma K, Nakayama O, Yamauchi T and Ando H **1997** *SAE* 970541.
Ipp W, Wagner H K, Wensing M, Leipertz A, Arndt S and Jain A K **1999** *SAE* 1999-01-0498.
Le Coz J F **1998** *9th Int. Symp. appl. Laser Tech. to Fluid Mechanics*, **1,** Paper 7.3.
Nouri J M, Brehm C and Whitelaw J H **1999** *ILASS99 Symposium*, 5-7 July, Toulouse, France.
Nouri J M and Whitelaw J H **2001** *Exp. Fluids*, **31**, 377-383.
Ortmann R, Arndt S, Raimann J, Grzeszik R and Wurfel G **2001** *SAE* 2001-01-0970.
Shelby M H, VanDerWege B A and Hochgreb S **1998** *SAE* 980160, 67-84.
Shrimpton J S, Yule A J, Akhtar P, Wigley G and Wagner T **1997** *Proc. ILASS-97*, pp 96-102, Florence, Italy.

Wigley G **1993** Optical diagnostics for flow processes. Edited by Lading et al.,175-204, (New York and London: Plenum Press).

Wigley G, Hargrave G K and Heath J **1998** *9th Int. Symp. Appl. Laser Tech. to Fluid Mechanics*, **1**, Paper 9.4.

Wirth M, Piock, W F, Fraidl G K K, Schoeggi P and Winklhofer E **1998** *SAE* 980492, 85-99.

Xu M and Markle L E **1998**. *SAE* 980493, 1998.

Yanta W J **1973** Naval Ordnance Laboratory, Report NOLTR 73-94, White Oak, Silver Spring, CA, USA.

Inst. Phys. Conf. Ser. No. 177
Paper presented at 1st Int. Conf. on Optical & Laser Diagnostics, London, 16–20 Dec. 2002
©2003 IOP Publishing Ltd

Quantitative characterisation of diesel sprays using digital imaging techniques

J Shao, Y Yan

Advanced Instrumentation and Control Research Centre, School of Engineering,
University of Greenwich at Medway, Chatham Maritime, Kent ME4 4TB, UK

G. Greeves, S Smith

Delphi Diesel Systems Ltd
Courteney Road, Hoath Way, Gillingham, Kent ME8 0RU, UK

Abstract: In recent years, research in diesel engine technology has been motivated by the desire to meet increasingly stringent emissions standards without unacceptable compromise of performance and efficiency. An important way to achieve this goal is to enhance the performance of the fuel injection system so that fuel/air mixing and combustion are made more complete. Unfortunately, the mechanism of fuel spray formation and the influence of fuel injection parameters on the resulting spray structure are still not very clear. It is therefore highly desirable to develop a reliable and efficient means for the quantitative analysis and characterisation of diesel sprays.

This paper presents the application of digital imaging and image processing techniques for the quantitative characterisation of diesel sprays. An optically accessible, constant volume chamber was configured to allow direct photographic imaging of diesel sprays. A high-resolution CCD camera and a flash light source were used to capture the images of the sprays. Dedicated image processing software has been developed to quantify a set of macroscopic, characteristic parameters of the sprays including tip penetration, near and far field angles. The spray parameters produced using the software are compared with those obtained using manual methods. The results obtained under typical spray conditions demonstrate that the software is capable of producing more accurate, consistent and efficient results than the manual methods. An application of the imaging processing software to the characterisation of diesel sprays for a VCO nozzle is also presented and discussed.

1. Introduction

In recent years, the diesel engine has become an extensively utilized power source choice in transportation and stationary systems due to its high power capacity and efficiency. Research in diesel engine technology has been motivated by the desire to meet increasingly stringent emissions standards without unacceptable compromise of performance. An important way to achieve this goal is to enhance the performance of the fuel injection system so that fuel/air mixing and combustion are made more complete [1]. Unfortunately, the mechanism of fuel spray formation and the influence of the operating parameters on the resulting spray structure are still not very clear. An in-depth understanding of the dynamic behaviour of the sprays under a high pressure environment is therefore essential.

Diesel spray analysis and characterisation have been ongoing for over fifty years through a variety of methods including mechanical, electrical and optical techniques. However, optical techniques are generally agreed upon as being the best choice, since they are relatively non-intrusive. Laser Beam Extinction, Particle Image Velocimetry, Phase Doppler Particle Analyser, Laser Doppler Velocimetry and Photographic Imaging [2-9] are all based upon optical techniques. Among these techniques, direct photographic imaging is particularly suitable for the quantification of the global characteristics of fuel sprays throughout the injection period. Traditionally, combustion engineers have to estimate the spray parameters from the images manually with the aid of a ruler and a protractor. Such a manual approach is obviously tedious and time-consuming and is thus unsuitable for the analysis and processing of a large number of spray images. Furthermore, accurate and detailed quantification of spray characteristics may not be possible due to the intrinsic limitations of the manual approach. It is for these reasons that automated, efficient and accurate processing of the spray images using digital imaging and image processing techniques has become highly desirable.

This paper focuses on the design, implementation and evaluation of digital image processing software developed for quantifying the fundamental characteristics of diesel sprays. An application of the software to the characterisation of diesel sprays is also presented.

2. System Set-Up

The system set-up used to capture spray images is shown schematically in Fig. 1. All tests were carried out in an optically accessible, constant volume chamber with a depth of 105 mm and a diameter of 63.5 mm under a non-evaporating and pressurized environment. A 40 mm thick quartz window was fixed to one end of the chamber to give excellent optical access. Delphi Diesel Common Rail fuel injection system was used to produce an injection pressure ranging between 150 and 1600 bar. The injector was mounted at another end of the chamber. The nozzle tested is a VCO type and has six holes each having a diameter (D) of 0.152 mm and a length (L) of 1 mm. A ring-collimated beam formed by a flash source was employed to illuminate the sprays. Digital images of the sprays were captured using a CCD camera having 1280×1024 pixels with an 8-bit grey scale resolution.

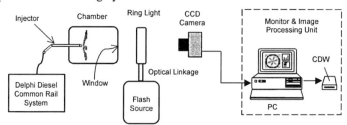

Fig.1 Schematic diagram of the system set-up

3. Definitions of Spray Parameters

The spray characteristics presented in this paper are represented in terms of three macroscopic parameters: tip penetration, near field angle and far field angle. Since detailed definitions of these parameters have been given elsewhere [9], only a brief description is presented here. Fig. 2 illustrates the definitions of these three parameters. The tip penetration is defined as the maximum distance between the tip and root of the spray. In diesel spray analysis, a parameter, called *spray angle*, is often used, which is defined as the angle between the tangents to the spray envelope. However, in order to achieve a better understanding of the dynamic properties of the spray, *near field angle* and *far field angle* are introduced, which are derived from the original concept of spray angle. As shown in Fig.2, the near field angle is measured between 60D and 100D from the nozzle tip while the far field angle from 100D to the far downstream of the nozzle tip.

4. Digital Image Processing Software

A logical image processing procedure has been established to derive the spray parameters from the raw spray images. Fig.3 shows the flowchart of the image processing procedure. The procedure can be subdivided into a number of intermediate steps. The following is a short description of the steps.

(a) Location of the reference point
The location of the spray plumes may vary slightly from one image to another for various reasons. It is therefore important to identify the exact location of the plumes in each image prior to segmentation. This is achieved by locating a reference point, which is defined as the geometrical centre of the nozzle in the background image captured prior to fuel injection into the chamber. Template matching was used to locate the reference points in the background image and spray images [9].

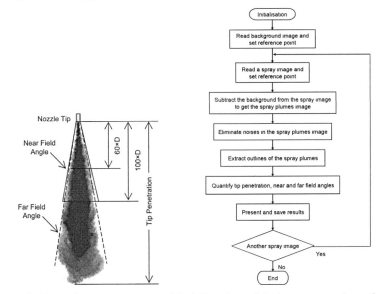

Fig.2 Definitions of spray parameters   Fig.3 Flowchart of the image processing software
(D: diameter of the nozzle hole)

(b) Removal of the background from the spray image
Once the reference point in each image is identified, the background is subtracted from the spray image. The resulting image contains mainly the spray plumes. However, impulsive or salt-and-pepper noise may remain in the subtracted image due to the noise of the imaging system. A threshold grey-level is then applied to the subtracted image to remove most of the background noise.

(c) Removal of the remaining scattered spots in the spray image
Despite the application of segmentation and threshold filtering, randomly scattered, 'messy' spots in the processed image remain due to the presence of random spray droplets. To remove these remaining unwanted clusters of pixels, a noise cut-off algorithm, i.e. *connected*

components labelling and sorting, is applied. In this study, an efficient run-length implementation of the local table method [10] was used to label all connected components in the spray plume image. Since the number of the spray plumes is known and the areas of the spray plumes are significantly larger than those of the unwanted connected components, spray plumes can be obtained by sorting the components and ignoring the smaller ones in the components queue.

(d) Determination of spray outline and calculation of spray parameters
An outline of the sprays is determined from the processed image by identifying the 'edges' of the spray plumes using edge detection technique. Sobel's edge detection method [11] was adopted in this study. The spray parameters as defined in Section 3 are then determined from the spray outline.

Digital image processing software incorporating the above algorithms has been developed. Fig.4 is a snapshot of the user interface.

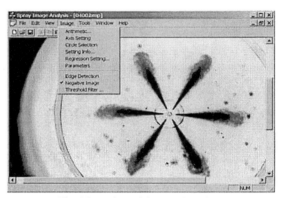

Fig.4 Snapshot of the user interface

5. Evaluation of the imaging processing software

It is necessary to evaluate if the software produces correct results for a given set of spray images. This is achieved by comparing the spray parameters produced by the software and those measured manually using a ruler and a protractor. A series of spray images was taken under typical spray conditions. The images were printed out on a laser printer for the manual measurement of spray parameters. The same set of images was then processed using the image analysis software. In order to avoid subjective errors the manual processing of the images was conducted before the software processing.

It is estimated that, without the consideration of printing distortion and human subjective errors, the resolutions of the manual method are ± 0.5 mm for tip penetration, $\pm 1°$ and $\pm 0.5°$ for near and far field angles, respectively. The resolutions of the software techniques are ± 0.06 mm, $\pm 0.4°$ and $\pm 0.2°$, respectively. Table 1 summarises the discrepancies between the results for the three parameters obtained using both the manual and software methods.

Table 1 Summary of comparison between the manual and software methods

	Absolute difference	Relative difference
Tip penetration	< 0.4 mm	$< \pm 1.5\%$
Near field angle	$< 1.2°$	$< \pm 6\%$
Far field angle	$< 1.0°$	$< \pm 5\%$

It is evident that the software has produced results that are at least comparable to those obtained using the manual approach. The main advantages of the software method over the manual approach can be summarised as follows:

(a) Efficiency: the software method is substantially more efficient than the manual approach. Automated software processing takes only a couple of minutes using a Pentium III 500MHz PC, whereas the manual approach requires at least a couple of days.
(b) Consistency: the software produces exactly same results for the same set of images under the same software configuration regardless of the operator and the time of operation. However, the same operator would likely produce different results at different times for the same images manually, let alone different operators.
(c) Accuracy: the automated method can process images with a higher resolution than the manual approach and will not be affected by printing distortion and human errors. The software should therefore produce more accurate results than the manual method.
(d) Flexibility: an operator can readily alter threshold-filtering parameters in the software to suit a broad range of spray conditions. In contrast, it is impossible for the manual method to maintain a consistency under such conditions.

6. Application of the Image Processing Software to the Characterisation of Diesel Sprays

To demonstrate the capabilities of the image processing software for the characterisation of diesel sprays under different conditions, a series of experiments was conducted on the set-up described in Section 2. Table 2 summarises the conditions under which the spray images were captured.

Table 2 Test conditions

Case	Injection pressure (bar)	Chamber pressure (bar)
1	600	25
2	600	40
3	1400	25
4	1400	40

By using the image processing software, variations of the three spray parameters with time were obtained. Since the six sprays exhibit similar dynamic attributes, only those of one of the sprays are presented in Figs.5-10 for technical clarity and succinctness.

Figs.5 and 6 show the variation of tip penetration with time under injection pressures of 600 bar and 1400 bar respectively. The tip penetration under a lower chamber pressure is longer than that under a higher chamber pressure for the same injection pressure. The tip penetration under a lower injection pressure is shorter than that under a higher injection pressure for the same chamber pressure. Figs.7-10 show the two kinds of spray angle for different chamber pressures and injection pressures. Fig.7 suggests that a higher chamber pressure entails a slightly larger near field angle. Fig.8 implies that a higher chamber pressure also leads to a larger far field angle. The near and far field angles shown in Figs.9 and 10 behave very similarly with those in Figs.7 and 8.

This application of the software has clearly demonstrated that the capability of the software for the quantitative characterisation of diesel sprays under different experimental conditions.

136

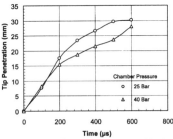

Fig.5 Variations of tip penetration with time
under injection pressure of 600 bar

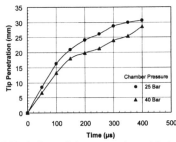

Fig.6 Variations of tip penetration with time
under injection pressure of 1400 bar

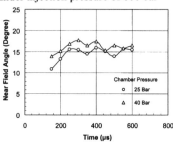

Fig.7 Variations of near field angle with time
under injection pressure of 600 bar

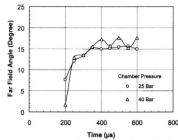

Fig.8 Variations of far field angle with time
under injection pressure of 600 bar

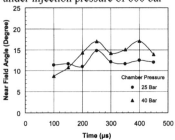

Fig.9 Variations of near field angle with time
under injection pressure of 1400 bar

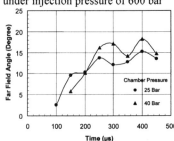

Fig.10 Variations of far field angle with time
under injection pressure of 1400 bar

7. Conclusions

The results presented have demonstrated that photographic imaging in conjunction with digital image processing algorithms is a powerful technique for the quantitative analysis and characterisation of the diesel spray process. The digital image processing software developed has played a vital part in achieving the automated quantification of the parameters of the sprays. The evaluation results of the image processing software have demonstrated its efficiency, accuracy, consistency and flexibility that cannot be achieved using the manual approach. The software has already been applied to characterise diesel sprays from a VCO nozzle under different conditions. It is envisaged that the software developed will play an important part in the in-depth understanding of the diesel spray process and subsequent optimisation of nozzles and diesel injection systems.

Acknowledgment

This paper is published by kind permission of Delphi Diesel Systems Ltd (UK) and the support provided by Delphi to carry out the work at University of Greenwich is acknowledged. Particular thanks are due to Hermann Breitbach, Chief Engineer, Advanced Engineering, for his support and comments and to Colin North for co-ordinating the project, both of Delphi Diesel Systems, Gillingham.

References

[1]   S. Tullis and G. Greeves, Seminar on Fuel Injection Systems, I.Mech.E., S492/S18, 1999.

[2]   L. Araneo, A. Coghe, G. Brunello and G. Cossali, SAE Paper No. 1999-01-0525, 1999.

[3]   K. Prescher, A. Astachow, G. Kruger and K. Hintze, SAE Paper No. 1999-01-0520, 1999.

[4]   H. Hiroyasu and M. Arai, SAE Paper No. 900475, 1990.

[5]   K. R. Brown, I. M. Partridge and G. Greeves, SAE Paper No 860233, 1986.

[6]   K. R. Kunkulagunta, SAE Paper No. 2000-01-1255, 2000.

[7]   M. Xu and H. Hiroyasu, SAE Paper No. 902077, 1990.

[8]   P. J. Tennison, T. L. Georjon, P. V. Farrell and R. D. Reitz, SAE Paper No. 980810, 1998.

[9]   J. Shao and Y. Yan, Sensors and their Applications XI, City University, London, Sep. 2001.

[10]   R. M. Haralick and L. G. Shapiro, *Computer and robot vision*, Addison-Wesley publishing company, 1992.

[11]   J. S. Lim, *Two-dimensional signal and image processing*, Prentice Hall, International Editions, 1990.

Inst. Phys. Conf. Ser. No. 177
Paper presented at 1st Int. Conf. on Optical & Laser Diagnostics, London, 16–20 Dec. 2002

Mobile Laser-Based Optical Triangulation System for the Measurement of Structures

S F Turner, D M Johnson

Laser Rail Ltd., Fitology House, Smedley Street East, Matlock, Derbyshire DE4 3GH

Abstract. Laser Rail have developed a new laser-based system for non-contact measurement of the internal profiles of railway structures at speed. Clearances are often tight and need to be checked every time the track through a tunnel or bridge is realigned. The measurement system works by optical triangulation: illumination of a full cross-section by laser, then measurement of the size of the illuminated profile by 9 digital video cameras.

The surface reflection characteristics of sooty tunnel walls are discussed together with solar intensity at different wavelengths, and safety considerations. These determine the type and arrangement of lasers. Image processing methods are outlined for identifying the narrow bright line of the illuminated profile, identifying the corners of the rail heads, relating the structure information to them and reducing the data into the closest profile in a given length of track.

A system of hydrophilic and hydrophobic polymer windows is described which allows the laser light to project upwards and outwards with protection but no distortion from raindrops, and allows the cameras to view the illuminated profile without interruption from a windscreen wiper.

Recorded tunnel profiles are presented and the system performance is briefly discussed.

1. Introduction

This paper describes a new laser-based system for non-contact measurement of the internal profiles of tunnels and other structures at speed. It has been developed for a railway application. Modern rail vehicles are larger than the Victorian vehicles for which Britain's rail network was built; they operate much faster and the latest tilting vehicles lean into curves. Modern vehicles therefore pass much closer to the lineside structures. To run the trains safely at high speeds, it is necessary to have accurate up-to-date measurements of the space available. Most structures are stable but the track settles and is realigned, every 3-6 months in some areas. Measurement of available clearance should ideally be performed at similar intervals.

This paper describes the technologies used in the system and design considerations for their use on the railway.

2. Background

There are several methods of measurement of a cross-sectional profile of a structure with respect to the rail head corners. The most simple is to use a long pole to hold one end of a tape measure on a point on the structure, measure the distance from this point to each of the running rails, then apply the cosine rule to give (x,y) coordinates for that point. This is a very slow method, particularly on an arched structure, and is inaccurate.

Quicker measurements can be made with scanning laser distance measurement devices, which detect time of flight or phase change on reflection. A static portable device is quite widely used on the railway, which scans a single profile wherever it is placed on the track, taking a few minutes to do so. Vehicle mounted scanner systems are also available but their disadvantage is that they map out a helix.

A third option is to use optical triangulation, demonstrated in Figure 1. A laser emits a beam of light in a fixed direction, and the position of the image on a detector can be used to give the distance from the laser to the object (Figure 1a). If the laser has a fanned beam, and a 2-D detector is used, the profile of an object can be measured (Figure 1b).

Moving the object or laser/camera system allows a series of profiles of the object to be measured. Sections are not missed, as they are by a scanning system. The speed of measurement is only limited by the frame speed of the camera and its integration time, which depends on the power of the reflected light. Objects which only appear for part of the integration time can also be detected if their reflectivity is sufficient.

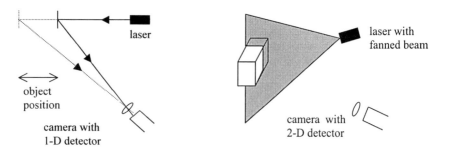

Figure 1a. Optical Triangulation in 1-D   Figure 1b. Optical Triangulation in 2-D

Optical triangulation has been used before on the railway, on the British Rail Structure Gauging Train [1]. This was built in the 1980s, using white light and television cameras, and has been used to measure structures for 20 years. Its speed and accuracy have been limited by the technology then available, but it successfully demonstrated the optical triangulation measurement principle in the target environment of the railway.

3. Design

Optical triangulation is the operational principle used for the Laser Gauging System. The method can be extended to measure a complete cross-sectional profile by using several laser-camera systems fitted all around the vehicle. In fact our prototype vehicle has 105 lasers and 9 high speed digital video cameras.

The arrangement of lasers and cameras was deduced from an investigation of the reflectivity of tunnel walls at different angles. Optical triangulation gives the best resolution per pixel at high angles of reflection, but reflectivity decreases at high angles, so there is a limit beyond which the image may be too dim for measurement.

Railway structures can be built of brick, concrete or stone, hewn out of rock, or fabricated from galvanized or painted metal. The reflectivity is normally determined by a thick, matt black surface layer of soot and diesel exhaust deposits. Measurements of diffusely reflected light have been published for rough oxidized metal from furnaces [2] and randomly rough gold and cobalt surfaces [3]. These papers report bright specular peaks in reflectivity then a broad plateau whose width can be linked to the roughness of the surface. Figure 2a gives our measurements of the reflectivity of tunnel walls at different angles. The measurements were taken using an Ophir PD200 laser power meter. A 22 mW output, 690 nm laser diode with a fan angle of 60° was used as an extended source, situated 115 mm from the wall surface, fixed at the 90° point of a modified protractor. A 7 mm square detector was moved around the protractor perimeter pointing at the illuminated line on the wall from a distance of 125 mm. The angles are accurate to +- 3°. The specular peak could not be measured in this way as the detector obscured the incident light, but our interest was in the higher angles. A continual slope was observed rather than a plateau, with gradient affected by surface.

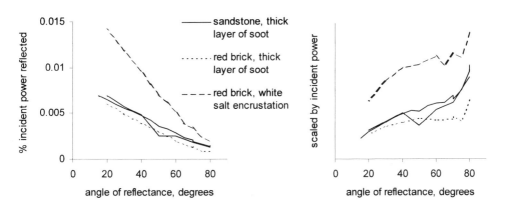

Figure 2. Reflectivity of tunnel walls with fixed laser source (a), and scaled to model the change in incident power implicit in change of reflectance angle in an optical triangulation system (b).

For the Laser Gauging System, large angles of reflection occur when the vehicle is close to a structure, which is when the incident laser power density is highest. The data of Figure 2a has been scaled by tan (angle of reflectance) to take this into account in Figure 2b. The image area on the prototype vehicle has been set so that the angle of reflection ranges from 30° to 72°, where the traces in Figure 2b are fairly constant.

A large number of low powered, visible laser diodes have been used to provide the illumination so that the power of each one falls within the Class II safety limits – where the eye is protected by its natural aversion responses such as the blink reflex. Solar irradiance is high within the visible part of the spectrum, but wavelength of 690 nm takes advantage of a decrease towards the infrared region [4]. Lasers within Class II limits cannot compete with direct sunlight, even with narrow band optical filters on the cameras, but this wavelength allows measurement of open structures in twilight or on dull days, as well as total darkness.

The cameras have been developed specifically for range imaging, and are programmed to detect a bright line through the image [5]. Their detection algorithm is demonstrated in Figure 3. A threshold is set according to the ambient light conditions. The processor scans across each row of photodiodes in turn and establishes the left edge and the right edge of the first band where this threshold is exceeded. If the bright band is over a certain width, it is rejected – this is likely to be the light at the end of the tunnel rather than the laser line on the tunnel wall. The scan is continued until a narrow bright band is found. The mid-pixel gives the centre of the bright band, with opportunity for sub-pixel resolution. Further on-chip image processing techniques allows processing of rows of photodiodes together to increase the signal-to-noise ratio and frame speed at the expense of resolution in one of the two dimensions.

For measurements on the railway, the optical system has to be exposed to the prevailing weather conditions. It is not a trivial issue to be able to project beams of laser light vertically upwards without distortion or water damage in heavy rain. At high speeds, it might be possible to project through a narrow slot which would collect rain and debris on its rear edge. The vehicle must be able to work at slow speeds, however, if narrow objects such as sign boards are to be detected.

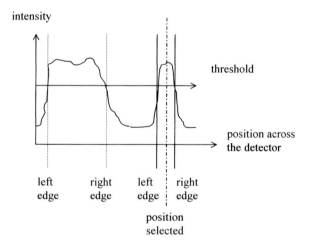

Figure 3. Detection of a narrow bright line by the camera.

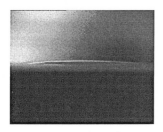

Figure 4a. Water droplet on a hydrophilic,   Figure 4b. Water droplet on a hydrophobic
etched and cleaned silicon surface. [6]   spun polystyrene surface. [6]

The prototype vehicle uses hydrophilic and hydrophobic polymer windows. Water droplets spread out over a hydrophilic surface, as in Figure 4a, but form more spherical droplets on a hydrophobic surface, as in Figure 4b.For the lasers, it is important to avoid a lens effect that would disperse parts of the line of the illuminated profile. Very thin hydrophilic plastic strips have therefore been used. To further encourage the water drops to spread out on the surface, we bend 0.1 mm x 70 mm x 1.8 m strips into a long half-cylinder, with the lasers mounted in a row under the highest part.

For the camera viewing windows, rotating panes with wipers mounted out of view were originally considered. These have been used on other systems [1]. Such complexity was found to be unnecessary, however. Circular water drops on a camera window do not distort the image significantly as the window is close to the camera lens. The droplets make slightly lighter circles on the camera image, but not enough to cause the laser line to be missed, or to fall on a different pixel. As all of the camera windows are steeply sloping, hydrophobic plastic was used so that the water droplets run off quickly.

4. Data Processing and Output

Each camera sends its data to two central computers. These determine the closest points encountered in radial bands of user-controlled width over 1m or 5m sections of track to reduce the data to manageable proportions. They also calibrate the camera images. The calibration shape consists of several rows of zigzag projections of different length, fitted all around and under the vehicle in the rail area. The real positions of the corners of each zigzag are known. The corners are automatically detected by the cameras as the vehicle passes each shape in turn. A 9-variable translation and rotation equation can then be derived for each camera to map the image pixel coordinates to the correct coordinates.

Once the images are calibrated, the rail head corner positions are identified, as shown by a cross in Figure 5b. A line across the track is extrapolated from the highest point on the top of the rail head to meet a perpendicular line from the most inward point on the side of the rail head. To avoid confusion from check-rails or over switches and crossings, the previous rail head corner position is used as a starting point, and the scanning is limited to 70 mm across and 35 mm down the rail head. The potential movement of the rail head corner position within the camera image is only ~10 mm over a 1m section of track.

The final output is the set of structure coordinates plotted relative to an origin at the left hand rail head corner, according the axes of true horizontal and vertical. A typical structure profile is shown in Figure 5a.

144

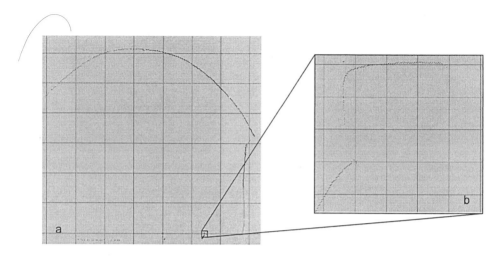

Figure 5. Cross-sectional profile inside the tunnel at the Churnet Valley Railway, marked in 1m squares (a), and an enlarged view showing the rail head corner identification marked in 20 mm squares (b).

5. Performance

All the above design principles have been implemented on a prototype road-rail Land Rover, shown in Figure 6.

Figure 6. Prototype road-rail land rover at Churnet Valley Railway

This vehicle is now fully functional and undergoing evaluation on private railways. It is hoped that it will be in use on Britain's rail network in the Autumn of 2002. Measurement accuracy varies between +- 5 and +- 10 mm throughout the images. The road-rail prototype has a top speed on the rails of 20mph, but it can be tested at higher speeds on the roads. The only limit to the operating speed of the system is the brightness of the lasers, which determines the integration time required and therefore the frequency of measurement along the track.

6. Conclusion

The design issues behind the production of Laser Rail's laser gauging system have been discussed. Optimum parameters are given for laser power and arrangement, camera viewing angles and protection. The image processing methods have been outlined. A prototype vehicle has been produced, and should be in use by Autumn 2002.

References

[1]   Edworthy M 1986 *Proceedings of the Society of Photo-Optical Instrumentation Engineers* **654** 35-42
[2]   Burnell J G, Nicholas J V and White D R 1995 *Optical Engineering* **34** no.6 1749-1755
[3]   McGurn A R *Surface Science Reports* 1990 **10** 357-410
[4]   McCartney E J *Optics of the Atmosphere* 1976 John Wiley
[5]   Johannesson M *Linköping Studies in Science and Technology Dissertations* 1995 **399**
[6]   Turner SF *Competitive Adsorption of Gelatin and Aqueous Surfactants* 1999 PhD Thesis, University of Cambridge

Inst. Phys. Conf. Ser. No. 177
Paper presented at 1st Int. Conf. on Optical & Laser Diagnostics, London, 16–20 Dec. 2002
©2003 IOP Publishing Ltd

The Interrogation of Multipoint Fibre Optic Sensor Signals Utilising Artificial Neural Network Pattern Recognition Techniques

William B Lyons, Colin Flanagan, Elfed Lewis

Department of Electronic and Computer Engineering,
College of Informatics and Electronics,
University of Limerick,
Limerick, Ireland.

Abstract. An optical fibre multipoint sensor system incorporating multiple (3) U-bend sensors is presented which is capable of detecting contaminants in water and depositions by coating on its surface. Interrogation of the data arising from the sensor is achieved using Artificial Neural Networks (ANN) pattern recognition. The system described is therefore capable of measurement at multiple points on a single fibre loop. It is also capable of recognising cross-sensitivity from interfering parameters such as lime scale coating in hard water and the presence of other species e.g. alcohol in the water. The signal analysis is performed using Artificial Neural Networks which allows classification of the samples under test, thus allowing the true measurand to be recognised.

1. Introduction

One of the main difficulties associated with Distributed or Multi-point fibre optic sensors, is the interference caused by the cross-coupling of signals from external parameters that can lead to difficulties in distinguishing the true measurand. A classical example of such cross-coupling interference is strain effects on a fibre optic temperature sensor and vice.

Since the level of complexity of the observed signals is high, it is necessary to address this problem using novel signal processing techniques. Neural Networks coupled with pattern recognition techniques provides such a novel solution in this case, as it is possible to train the networks to recognise characteristic features in the OTDR signals arising from the fibre sensors.

It has been proposed that for many applications of optical fibre multipoint sensors, artificial neural network pattern recognition techniques may be used to resolve the problems arising from cross-sensitivity to other parameters [1]. Previous work by Lyons *et al* [2], in which data from a single U-bend optical fibre sensor was used initially to train a single layer perceptron and then a multi layer perceptron, and a comparative analysis of their results showed that a multi-layer perceptron was required to adequately classify the

148

data. This provided a means of validating the technique, which may be extended to the interpretation of more detailed data from similar sensors.

2. Experimental Setup

The sensor system – as illustrated in Figure 1 – consisted a continuous1 km length of 62.5 μm Plastic Clad Silica (PCS) fibre connected to an Optical Time Domain Reflectometer (OTDR), and a computer used for acquisition and analysis.

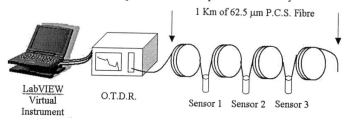

Figure 1. Measurement system configuration.

In order to minimise optical power loss associated with each sensor, and therefore increase the dynamic range of the sensing system, the sensors were fabricated from the same fibre as used to transmit light through the system (i.e. 62.5 μm PCS), hence removing the need to splice the three sensors into the transmitting fibre and any resulting associated power losses.

The sensitivity of the sensors was maximised by utilising an exposed core U-bend configuration where the cladding was removed and the core exposed directly to the absorbing fluid under test. The operation of these sensors is based on the modulation of the light intensity propagating in the fibre by the measurand as a result of the interaction with the evanescent field penetrating into the absorbing measurand. Much experimental work has already been reported [3-5] for a single U-bend sensor detailing the resulting sensitivity gains from the evanescent wave increases resulting from the curving of the sensing fibre [6].

2.1   Sensor Fabrication

A 2 cm length of buffer and cladding located at each of the three sensor locations (at 665m, 756m, and 857m from the launch end of the fibre) was chemically removed. After the exposed fibre regions were cleaned with acetone, they were slowly bent into a U-shape using heat from a flame. The bending procedure was controlled using an in-house developed fixture to improve the repeatability of the sensor manufacturing process and hence improve the reproducibility characteristics between successive sensors. The final bend radii were measured to be 1mm.

2.2   System Configuration

As illustrated in Figure 1, the system configuration comprised the optical fibre, the three U-bend sensor sections, an EXFO IQ7000 (0.85μm) OTDR and a Pentium III 800 MHz PC configured as a virtual instrument (VI) for data capture and pre-processing using a

LabVIEW. The LabVIEW VI programs were developed in house specifically for this investigation, and the resulting data output were made available for analysis by the artificial neural network implemented using SNNS V3.2 (Stuttgart Neural Network Simulator) [7].

A sample measurement trace obtained from the OTDR for all three sensors exposed to air (dry condition) is illustrated in Figure 2. Both the OTDR/fibre connector and Fresnel reflections can be seen on the left and right hand sides respectively. The three centre peaks are due to the U-bend sensor 1, sensor 2 and sensor 3 at 665m, 756m and 857m respectively. As with all multi-point or distributed sensor systems, it is important that the attenuation imposed by the launch optics and sensors nearest the launching end of the fibre do not impose too much attenuation on the transmitted signal as this would ultimately render the succeeding sensors useless. These losses have been minimised it this system and this is evident in Figure 2.

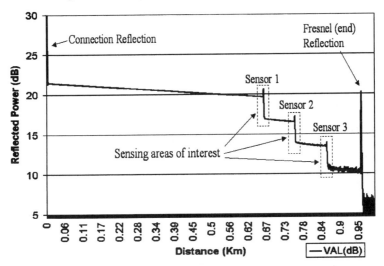

Figure 2. Sample OTDR trace from fibre indicating a dry sensor response for all three sensors.

3. Results

Numerous OTDR readings were taken for the various sensing conditions at each sensor simultaneously, i.e. the sensor systems response to response to various combinations of Water, 50% Ethanol, and Air present at the sensors. The various combinations are shown in Table 1. Also of interest were the resulting attenuation effects on the succeeding sensors for each of the combinations.

Large quantities of OTDR readings are required to produce a sufficiently large number of pattern files that are required to both successfully train and enable testing of an Artificial Neural Network (ANN).

TABLE 1: Sensor exposure combinations investigated.

Sensor 1	Sensor 2	Sensor 3	No. of Readings
Air	Ethanol	Water	51
Water	Air	Ethanol	51
Ethanol	Water	Air	51
Air	Water	Ethanol	51
Water	Ethanol	Air	51
Ethanol	Air	Water	51

4. Analysis

The overall OTDR trace – as illustrated in Figure 2 – consisted of 12000 data points, and since the sensing areas of interest (highlighted by dashed boxes in Figure 2) form a relatively small part of the over all trace (253 points – 91 for both sensor 1 and 2, and 71 for sensor 3), a feature extraction technique was carried out using an in-house designed VI that would locate the data peaks associated with the sensors, select the required window width and save this data from further pre-processing prior to application to the Stuttgart Neural Network Simulator (SNNS).

The further pre-processing involved the selected OTDR sensor data of interest being normalised between ±1 using the standard LabVIEW Scale 2D array VI [7] and then generated into the correctly formatted neural network pattern files. Figure 4 illustrates two examples of the resulting normalised data that was used to train the neural network – one plot shows the result of a 50% Ethanol, air, and distilled water exposure on Sensors 1, 2, and 3, respectively, and the other illustrates the result of an air, 50% Ethanol, and distilled water exposure on Sensors 1, 2, and 3, respectively.

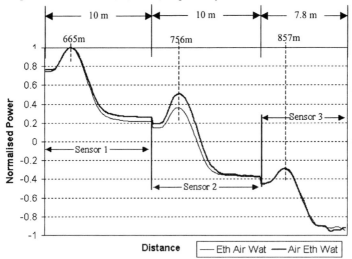

Figure 3. A comparison of a normalised 3 sensor data trace for 50% Ethanol/Air/Water and Air/50% Ethanol/Water at sensors 1, 2, and 3 respectively

It can be seen from the two examples in this illustration that not only is there sufficient difference between the readings to distinguish between the measurands, but also that the resulting attenuation at each sensor affects the attenuation (and pattern) of the succeeding sensors.

It is important to document that when dealing with an OTDR addressed multipoint sensor system (as in this investigation), any measurand induced changes in the power levels at a sensor (due to the resulting attenuation change) will have a direct effect on the shape in succeeding sensors on the fibre loop as a result of the remaining power left in the fibre to address the remaining sensors. The use of artificial neural networks allows this problem to be separated out, and give the correct reading for each sensor.

Using SNNS, a feed-forward three-layer neural network was constructed with the number of input nodes representing the number of points required to represent the three sensors response and attenuation (in this case 253). The output layer consisted of nine nodes – three per sensor, with each group of three representing a sensing condition under investigation. In order to determine the amount of nodes to be included in the hidden layer, five different feed-forward networks were examined with forty, thirty, twenty, ten and five nodes in the hidden layers respectively. It was found that a hidden layer of thirty nodes performed the best. An illustrative representation of the artificial neural network used – including the representation of the output nodes is given in Figure 4.

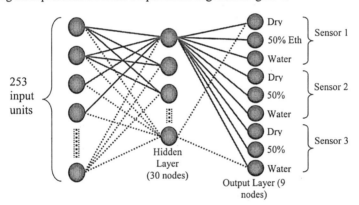

Figure 4. Illustration of feed-forward network constructed in SNNS including representation of the output nodes in the output layer.

A total of 270 result patterns were used to train the network – 45 for each condition listed previously in Table 1. For training the feed-forward network, 500 epochs were required with a back-propagation algorithm utilising a learning rate of 0.3. The network was initialised with randomised weights and trained with a 'topological order' update function.

In order to test the operation of the trained network, an independent set of data to that which had been applied in the training of the network was used. This was generated from the remaining unused 6 result patterns for each of the same conditions listed in Table 1. The resulting 36 patterns were applied to the trained network and were all classified correctly. Table 2 shows a sample of the results obtained when the test pattern set was applied to the trained network.

TABLE 2: Sample of 6 results from the Trained ANN Test Set.

Test Condition			Expected output			Observed output		
S1.	S2.	S3.	S1.	S2.	S3.	Sensor 1.	Sensor 2.	Sensor 3.
Air Eth Wat			1 0 0	0 1 0	0 0 1	0.990 0.002 0.036	0.001 0.964 0.035	0.023 0.018 0.962
Wat Eth Air			0 0 1	0 1 0	1 0 0	0.046 0.061 0.916	0.001 0.945 0.076	0.993 0.000 0.008
Wat Air Eth			0 0 1	1 0 0	0 1 0	0.013 0.069 0.928	0.999 0.000 0.013	0.000 0.930 0.055
Eth Wat Air			0 1 0	0 0 1	1 0 0	0.003 0.913 0.088	0.063 0.040 0.911	0.927 0.000 0.050
Eth Air Wat			0 1 0	1 0 0	0 0 1	0.000 0.949 0.052	0.999 0.000 0.008	0.001 0.042 0.972
Air Wat Eth			1 0 0	0 0 1	0 1 0	1.000 0.000 0.093	0.006 0.018 0.989	0.000 0.999 0.014

5. Conclusion

A multipoint optical fibre sensor system has been developed and successfully tested on a 1km length of optical fibre. The system is capable of detecting the presence of ethanol in water as well as detecting dry air conditions on each of three separate and independent sensors.

The sensor geometry chosen for this investigation was to have the fibre in a 'U-bend' configuration based on a 62.5 nm fibre. This provided sufficient sensitivity and introduced an acceptably small amount of loss per sensor through the system. The spatial resolution limit of the system was that of the OTDR which in this case is 1mm. Therefore the sensors could be located to this accuracy on the 1km fibre. A further improvement in the resolution of the dynamic range of the sensor elements is also possible but this was restricted in this investigation to simply detecting the presence of the contaminant in the water or if the sensor had become dry. Further developments of this system will allow degrees of contamination to be detected which will correspond to a greater number of output nodes in the network.

References

[1] Lyons W B and Lewis E 2000 *Trans. of the Inst. of Measurement and Control* **22** 385-404
[2] Lyons W B, Ewald H, Flanagan C, and Lewis E 2000 *Proc. Artificial Neural Networks in Engineering Conference 2000 (ANNIE 2000)* **10** 663-70
[3] Khijwania S K and Gupta B D 1998 *Optics Communications* **152** 259-62
[4] Khijwania S K and Gupta B D 1999 *Optical and Quantum Electronics* **31** 625-36
[5] Khijwania S K and Gupta B D 2000 *Optics Communications* **175**135-7
[6] Soichi Otsuki 1998 *Sensors and Actuators* **B53** 91-6
[7] Stuttgart Neural Network Simulator (SNNS) 1995 User Manual Ver 4.1 Rep.No. 6/95

Inst. Phys. Conf. Ser. No. 177
Paper presented at 1st Int. Conf. on Optical & Laser Diagnostics, London, 16–20 Dec. 2002
©2003 IOP Publishing Ltd

Fibre Bragg Grating coupled fluorescent optical fibre sensors

T Sun, D I Forsyth, S A Wade, S Pal, J Mandal and K T V Grattan
School of Engineering and Mathematical Sciences, City University, Northampton Square, London EC1V 0HB

W D N Pritchard
Corus Research Development & Technology, Corus Ltd, West Glamorgan SA13 2NG

G Garnham
BNFL Ltd, Lancashire PR4 OXJ

Abstract: Bragg gratings written in photosensitive rare earth doped fibres, using fluorescence decay time and wavelength shift interrogation have been investigated for their extended capability to address strain and temperature measurement simultaneously.

1. Introduction

Fibre Bragg gratings (FBGs) in optical fibres have demonstrated a wide range of applications, such as sensors, dispersion compensators and laser mirrors [1]. In particular, simultaneous strain and temperature monitoring is important for a variety of uses, especially in structural integrity determination and optical diagnostics [2]. Thermal characterization of the Fibre Bragg gratings is essential for increased reliability and lifetime of these components in wavelength division multiplexing (WDM) systems and for high temperature sensing applications [3], for example, oil well monitoring [4] and smart structures [5]. Recent work on material optimization reported by the University of Southampton has shown that tin-doped fibres are potential candidates for many applications in both telecommunication and sensors markets [6][7][8].

The exploration of an intrinsic dual sensor based on FBGs written in rare-earth doped fibres over a wide sensing range is the focus of this paper. The aim is for a device for high temperature structural monitoring and the system developed for the measurements uses a luminescent 'decay time' temperature compensated Bragg grating. The long term thermal characteristics of the FBGs written in rare-earth doped fibres as well as in Ge-B co-doped fibres have been investigated and further work will be carried out in varying the writing laser parameter as well as the nature of the photosensitive fibres used, to optimise the intrinsic dual sensor. This work builds upon previous reports by the authors [9] of the response of the temperature – sensitive characteristics of the system to these high temperatures – this paper focuses upon the capability of the grating part of the sensor to respond under the extreme conditions set by the industrial collaborators on the project.

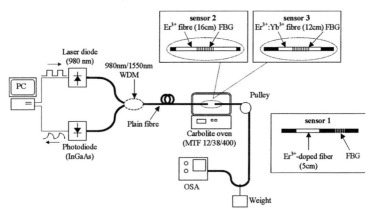

Figure 1 Dual sensor for strain and temperature measurement

2. Dual sensor for strain and temperature measurement

Three different types of sensor configurations have been tested, shown in Figure 1, designed for strain and temperature measurement by multiplexing the rare-earth doped fibres with fibre Bragg gratings. Sensor 1 shows a Bragg grating in a plain fibre attached to the temperature compensating element, and sensors 2 and 3 are constructed with gratings written into the doped fibre itself. To the knowledge of the authors, this is a unique application of Bragg gratings in sensors. Using these types of configuration, in this way the temperature and strain sensors are easily co-located and investigated in the fibre. These sensors have been tested and the results are summarized in Figure 2. The results show that the combination of the fibre grating and the luminescent doped fibre works well, with the measurement mechanisms being complementary, not competitive. More importantly, the gratings written into the doped fibres show the same strain response as a grating written in a plain (telecoms) fibre and thus the ability to use the fibre for luminescent decay time temperature-compensating measurements is unaffected.

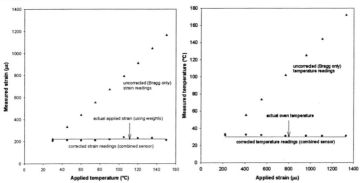

Figure 2 Strain or temperature compensation using the sensors in Figure 1

3. High temperature strain monitoring: thermal characteristics of FBGs written in silica fibres

This section of the work concerns the temperature stability of the FBGs used for the strain measurement. The luminescent fibre is stable to 'decay time' temperature measurement to the upper temperature limit quoted earlier, but previous work on the use of Bragg gratings at very high temperatures has been limited, as 'telecoms' applications rarely demand applications at temperatures anywhere approaching this range. Tests were carried out on the grating used in sensor 3 and results on the grating reflectivity with the gradual heating of the combined sensor are shown in Figure 3.

3.1 FBGs written in Er/Yb co-doped fibres

The grating written into sensor 3 was a type I grating produced using a double frequency argon ion laser centred at 244 nm with output power 100 mW. The co-doped fiber was of unique construction in that it had been designed with a photosensitive ring (during the manufacturing process). This fiber had a diameter of 125µm, a numerical aperture of 0.2 and core dopants of Er^{3+}, Yb^{3+} and aluminium. Figure 3 below shows the results of a 'step-wise' increase in temperature from one stable temperature to another, in 50 degree increments. The reflectivity of the grating was measured using the set-up shown in Figure 1, and it can be seen that at each stable temperature, the diminution of the reflectivity follows the single exponential function and the associated time for each is listed in the brackets. With the increase in the temperature, the decreased decay time of the reflectivity against temperature indicates faster aging process under higher temperatures. Clearly, beyond 400 degrees the grating reflectivity falls to a very low level and this makes this scheme essentially unsuited to measurement at temperatures higher than this, but below the upper temperature limit for the measurement set by the application. The results compare well with those reported by Brambilla [6][7].

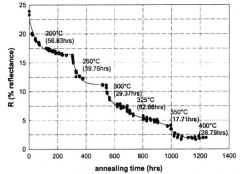

Figure 3 Temperature effects upon the grating reflectivity in sensor 3

3.2 FBGs written in photosensitive fibres

The setup for producing these gratings is shown in Figure 4, where the pulse power of the excimer laser (LAMBDAPHYSIK COMPEX 205) centred at 248 nm was adjusted to 18 mJ and the repetition rate to 15 Hz (in our latest setup, the above laser has been replaced by a new BraggStar 500 supplied by TuiLaser AG). The photosensitive fibre used here was supplied by Fibercore Ltd., the core of which is boron-germanium co-doped and the fibre was exposed to the UV beam for 30 seconds. The reflection spectra of the FBG thus produced, when subjected to different temperatures, are shown in Figure 5(a) where the higher temperature has reduced the reflectivity and shifted the peak wavelength of the grating. Figure 5(b) is the result of long term thermal stability tests on the grating produced and similar results to those shown previously from the FBG in Er/Yb co-doped fibres have been obtained. Again, the reflectivity diminishes in an exponential fashion at each stable temperature and when exposed to higher temperature, the grating decays faster. It is interesting to see from Figure 6 that although the reflectivity may change at the same constant temperature, the peak wavelength of the grating remains constant which further confirms the advantage of the wavelength shift based sensing scheme, based on the use of FBGs.

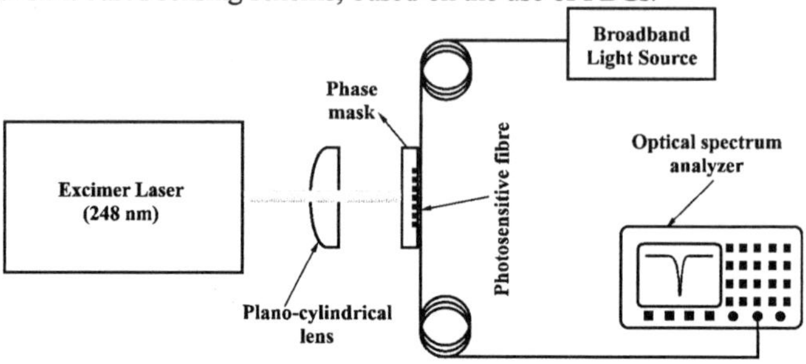

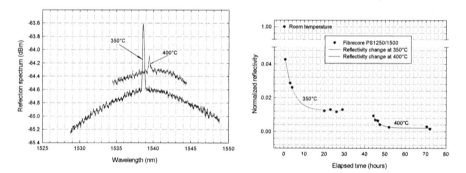

Figure 4 Experimental setup for FBG writing for this work

Figure 5(a) Reflection spectrum of the FBG fabricated by using a pulsed excimer laser; 5(b) Long term thermal stability tests of the FBG written in photosensitive fibres

3.3 FBGs written in various Er doped photosensitive fibres
The potential to increase the sensing range for a fluorescence-based dual (strain and temperature) sensor exists when FBGs are incorporated in rare-earth doped fibres: however the upper limit of the high temperature of the FBGs is related to the type of gratings written as well as the core composition [10] of the photosensitive fibres. Continuous work is being carried out to test the high temperature reliabilities of these different types of gratings and the results will be reported in the near future.

4. Summary

The work done has shown both the potential of the system illustrated for high temperature compensated strain-temperature measurements using co-located Bragg gratings in doped optical fibres and the need for further research work to extend the capability of the sensor for dual measurement up to the desired 700 degrees. Work is continuing to expand the sensing range of the fluorescence based optical fibre sensor coupled with FBGs to address strain and temperature simultaneously.

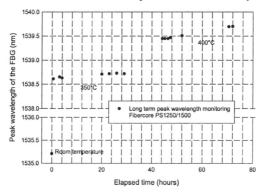

Figure 6 Long term characteristics of the peak wavelengths of the FBG

References

[1] A. Othonos, Bragg gratings in optical fibers: fundamentals and applications in *Optical Fiber Sensor Technology, Advanced Applications – Bragg Gratings and Distributed Sensors* (Eds. K.T.V.Grattan and B.T.Meggitt), Kluwer Academic Publishers, pp79-188, 2000

[2] J. D. C. Jones, Review of Fibre Sensor Techniques for Temperature-Strain Discrimination in *Proceedings of the 12th International Conference on Optical Fiber Sensors* (OSA Tech. Digest Series 16, Williamsburg, 1997), pp. 36-9, 1997

[3] J. Canning, M. Englund, K. Sommer, Fibre gratings for high temperature sensor applications, in *Proceedings of the 14th International Conference on Optical Fiber Sensors*, Postdeadline papers, pp. 1-4, 2000

[4] R. J. Schroeder, T. Yamate, E. Udd, High temperature and temperature sensing for the oil industry using fiber Bragg gratings written into side hole single mode fiber, Proceedings of SPIE – the International Society for Optical Engineering, Vol. 3746, pp42-5, 1999

[5] T. Wavering, J. Greene, S. Meller, T. Bailey, C. Kozikowski, S. Lenahan, K. Murphy, M. Camden, L. Simmons, High temperature optical fiber sensors for characterization of advanced composite aerospace materials, Proceedings of SPIE – the International Society for Optical Engineering, Vol. 3538, pp58-63, 1999

[6] G. Brambilla, H. J. Booth, E. K. Illy, L. Reekie, H. Rutt and L. Dong, Strong fibre Bragg-gratings written in highly photosensitive Sn-doped fibres without H$_2$-loading using KrF excimer and UV copper vapour lasers, Abstracts and Progamme of In-fibre Bragg gratings and special fibres, The Institute of Physics, 17th October 2001

[7] G. Brambilla, T. P. Newson and H. Rutt, Material optimisation for hight-temperature grating devices written by KrF excimer lasers, Abstracts and Progamme of In-fibre Bragg gratings and special fibres, The Institute of Physics, 17th October 2001

158

[8] D. Razafimahatratra, P. Niay, M. Douay, B. Poumellec and I. Riant, Comparison of isochronal and isothermal decays of Bragg gratings written through continuous-wave exposure of an unloaded germanosilicate fiber, Applied Optics, Vol.39, No.12. pp1924-33, 2000

[9] K. T. V. Grattan, Z. Y. Zhang, *Fibre Optic Fluorescent Thermometry*, Chapman & Hall, London, 1995

[10] R. Kashyap, Fiber Bragg Gratings, San Diego London, Academic Press, 1999

Inst. Phys. Conf. Ser. No. 177
Paper presented at 1st Int. Conf. on Optical & Laser Diagnostics, London, 16–20 Dec. 2002
©*2003 IOP Publishing Ltd*

Phase Response of Nonlinear Medium Induced by Transitions Between Excited Energy Levels

Jihad S M Addasi

Department of Basic Sciences - Tafila Applied University College, Al – Balqa'

Applied University

P.O. Box 179, Tafila-66110, Jordan.

Abstract. The light-induced change in the refractive index of dye solution and nonlinear effects are interested for development of optical information processing method. The nonlinear medium is modeled by three-level model (S0-S1-S2), where the light beam with intensity I12 at frequency $\omega 0$ is used to transmit the molecules into the first excited energy level (S1). Then waves of intensity I23 at frequency ω are used in the nonlinear process in the excited channel (S1-S2).

Control of nonlinear processes can be realized by independent light beam, not needed to be highly coherent or monochromatic. Taking into consideration the nonlinear properties for dye solution it has been found, that the principal (excited) channel is bleaching at radiation intensity I12(I23) decreasing with increasing of radiation intensity I23 (I12) in other channel.

1. Introduction

Activation of the molecules to the excited state by optical pumping and realization of nonlinear effects conditioned by the transitions between excited energy levels could be interesting to develop optical information processing methods [8]. Control over efficiency of nonlinear processes can be realized by using an independent light beam. This light beam in the general case needs not be highly coherent or monochromatic. In [1] self-diffraction of radiation in Rhodamine 6G dye was realized owing to absorption from the excited singlet molecular level ($\lambda=1.06\mu m$). In addition the dependence of self-diffraction efficiency on the intensity of optical pumping into the principal absorption band ($\lambda=0.53\mu m$) was demonstrated [1].

The light-induced change in the refractive index (phase response) of nonlinear medium and other nonlinear processes are studied by using two, three, four and five-level schemes [2, 4, 5]. The thermal nonlinearly for all channels (principal and excited) are taken into consideration along with a resonance nonlinearly at a frequency of waves involved in the nonlinear process due to week spectral selectivity of thermal nonlinearity [4, 5, 7].

2. Theory

A three-level scheme (S_0-S_1-S_2) is studied widely for nonlinear medium modulation, especially this model gives ability to use radiations with different frequencies [2, 4, 5]. The three-level model of nonlinear medium, demonstrated in figure 1, is used widely for dye solutions. According to this model the independent beam with intensity I_{12} at frequency ω_0 is tuned into the absorption band of the principal singlet-singlet transition (S_0-S_1) to translate the molecules into first excited energy level (S_1). In addition, the waves involved in the nonlinear process, with intensity I_{23} at frequency ω, realize the transitions in the excited singlet-singlet (S_1-S_2) channel. For this case, the solution of kinetic equations for the energy levels N_i populations in the stationary interaction mode may be written as follows:

$$N_1 = N\ (1+B_{21}I_{12}/vp_{21})(1+B_{32}I_{23}/vp_{32})/K\ ;$$

$$N_2 = N\ (B_{12}I_{12}/vp_{21})(1+B_{32}I_{23}/vp_{32})/K\ ; \qquad (1)$$

$$N_3 = N\ (B_{12}B_{23}I_{12}I_{23}/v^2p_{21}\ p_{32})/K\ ;$$

where

$K=1+(B_{12}+B_{21})I_{12}/vP_{21}+B_{32}I_{23}/vP_{32}+(B_{12}B_{23}+(B_{12}+B_{21})B_{32})I_{12}I_{23}/v^2P_{21}P_{32}$; $N=N_1+N_2+N_3$;

N - is the number of molecules in the unit volume of nonlinear medium; P_{ij} - is the total

probability of spontaneous and radiationless transitions in the i-j channel; v=c/n - is the light

velocity in the nonlinear medium. The Einstein coefficients B_{12}; B_{21} - are determined at the

pumping frequency ω_0, at the same time B_{23}; B_{32} - are determined at waves frequency ω in

excited channel.

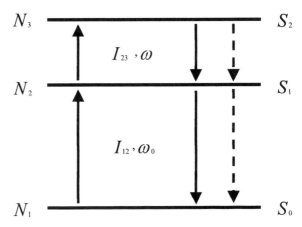

Figure 1 Diagram of three-level model. The solid lines denote the radiation-induced transitions of

molecules, and the dashed lines denote spontaneous and radiationless transitions. Where: S_0 - is the

ground state; S_1 (S_2) - are the first (second) excited states; N_i - is the population of i- energy level;

I_{12}, ω_0 (I_{23}, ω) - are the intensity, frequency of radiation in principal (excited) channel.

The complex refractive index of nonlinear medium $\hat{n}_{ij} = n_{ij} + i\chi_{ij}$ due to resonance

transitions in i-j channel can be determined [5] by the following equation:

$$\hat{n}_{ij}(\omega) = \frac{\hbar c}{2v}(N_i\hat{\theta}_{ij}(\omega) - N_j\hat{\theta}_{ji}(\omega)) ; \tag{2}$$

where

ω – is the frequency of radiations acting in i-j channel. $\hat{\theta}_{ij}(\omega) = \theta_{ij}(\omega) + iB_{ij}(\omega)$; $\theta_{ij}(\omega)$ -

are related by the dispersion relations to the Einstein coefficients $B_{ij}(\omega)$.

By using equations (1) and (2) the complex refractive index of nonlinear medium due to the transitions in the excited singlet-singlet (S_1-S_2) channel can be written in the following form:

$$\hat{n}_{23} = \chi_0 (I_{12}/vp_{21})(\hat{\theta}_{23}(1 + B_{32}I_{23}/vp_{32}) - \hat{\theta}_{32}B_{23}I_{23}/vp_{32})/K \quad ; \qquad (3)$$

where

$\chi_0 = N\hbar c B_{12}/2v$ - is the linear extinction coefficient. For the intensity $I_{23}=0$, the refractive index $\hat{n}_{23}$ has its initial value:

$$\hat{n}_{23}^0 = \chi_0 \hat{\theta}_{23}(I_{12}/vp_{21})/(1 + (B_{12}+B_{21})I_{12}/vp_{21}). \qquad (4)$$

Consequently, a change in the refractive index ($\Delta\hat{n}_{23} = \hat{n}_{23} - \hat{n}_{23}^0 = \Delta n_{23} + i\Delta\chi_{23}$) due to transitions in the excited channel, induced by intensity I_{23}, is as follows:

$$\Delta\hat{n}_{23} = -\chi_0 (\frac{B_{23}I_{12}I_{23}}{v^2 p_{21} p_{32}}) \frac{\hat{\theta}_{23}B_{12}I_{12}/vp_{21} + \hat{\theta}_{32}(1 + (B_{12}+B_{21})I_{12}/vp_{21})}{(1 + (B_{12}+B_{21})I_{12}/vp_{21})K} \qquad (5)$$

The thermal change of the refractive index due to radiationless transitions in i-j channel under conditions of adiabatic heating of the nonlinear medium can be determined by the following equation:

$$\Delta n_{ij}^T = N_j(1 - \mu_{ji})p_{ji}\hbar\omega t (dn/dT)/C_\rho; \qquad (6)$$

where

μ_{ji} - is the quantum yield of luminescence; C_ρ – is heat capacity of the unit volume; dn/dT – is the thermooptic coefficient; t – is the time. Note that the radiation absorption in the excited channel (S_1-S_2), changing the population of the first excited level (S_1), results in a change of heat release in the principal (S_0-S_1) channel. Unlike the resonance phase response, the thermal variation of the refractive index due to its week spectral selectivity is determined by thermal release in both spectral channels [7]. Taking into consideration equations (1) and (6) the thermal change in the refractive index, in principal (S_0-S_1) and excited (S_1-S_2) channels, can be written by the following expressions:

$$\Delta n^T = \Delta n^T_{12} + \Delta n^T_{23};$$

$$\Delta n^T_{12} = -\chi_0 \left(\frac{B23 I_{12} I_{23}}{v^2 P_{21} P_{32}} \right) \frac{\sigma_{12}(1 - \mu_{21}) B_{12} I_{12}}{(1 + (B_{12} + B_{21}) I_{12}/vp_{21}) K}; \qquad (7)$$

$$\Delta n^T_{23} = \chi_0 \left(\frac{B23 I_{12} I_{23}}{v^2 P_{21} P_{32}} \right) \frac{\sigma_{23}(1 - \mu_{32})(vp_{32})(1 + (B_{12} + B_{21}) I_{12}/vp_{21})}{(1 + (B_{12} + B_{21}) I_{12}/vp_{21}) K};$$

where

$\sigma_{12} = 2\omega_0 (dn/dT) t / cC_p$; $\sigma_{23} = 2\omega (dn/dT) t / cC_p$. For consideration of heat release in

principal channel, is taken the portion of thermal refractive index (Δn^T_{12}, equation (7))

associated with transitions between the excited energy levels.

Analysis of equations (5) and (7), shows that the resonance ($\Delta n_{23}; \Delta\chi_{23}$) and thermal

($\Delta n^T_{12}; \Delta n^T_{23}$) changes in the refractive index take their saturation at the radiation intensity in

the excited channel:

$$I^{sat}_{23} = \frac{(1 + (B_{12} + B_{21}) I_{12}/vp_{21}) vp_{32}}{B_{32} + (B_{12} B_{23} + (B_{12} + B_{21}) B_{32}) I_{12}/vp_{21}}, \qquad (8)$$

for which all components of phase response reach one half their maximum value. The

saturation intensity in excited channel (I^{sat}_{23}) has its maximum value $I^{sat}_{23} = vP_{32}/B_{32}$ for

pumping beam intensity $I_{12} \approx 0$, and has its minimum value $I^{sat}_{23} = vP_{32}/(B_{23} + B_{32})$ for

effective population of first excited energy level ($I_{12} \gg vP_{21}/(B_{12}+B_{21})$; $B_{12} \gg B_{21}$).

The dependence of the phase response on the radiation intensity in principal channel (S_0-S_1)

shows that the resonance ($\Delta n_{23}; \Delta\chi_{23}$) and the thermal ($\Delta n^T_{12}; \Delta n^T_{23}$) changes in refractive

index are saturated at saturation intensity I^{sat}_{12} in principal channel with value:

$$I_{12}^{sat} = \frac{(1 + B_{32}I_{23}/vp_{32})vp_{21}}{B_{12} + B_{21} + (B_{12}B_{23} + (B_{12} + B_{21})B_{32})I_{23}/vp_{32}}, \tag{9}$$

when the pumping beam with intensity I_{12} tuned into the centre of absorption band in the principal channel ($\omega_0 = \omega_{12}$) and the radiations with intensity I_{23} tuned in the centre of the absorption band in the excited channel ($\omega = \omega_{23}$).

3. Analysis

The phase response of nonlinear medium depends on not only spectral parameters of nonlinear medium (B_{ij}, P_{ji}, σ_{ij}, etc.), but also on intensities and frequencies of pumping beam and of waves involved nonlinear processes (I_{12}, I_{23}, ω_0, ω). There are many applications on ethanol solution of Rhodamine 6G [1, 2, 4, 5], for which the radiation with wavelength $\lambda_0 = 0.53\mu m$ can be tuned in the centre of absorption band in principal channel. At the same time, the radiation with wavelength $\lambda = 1.06\mu m$ can be tuned in the centre of absorption band in excited channel of Rhodamine 6G. For that, the numerical analysis for equations (5) and (7) is taken for the nonlinear medium with the Gaussian form of mirror-symmetric absorption and emission bands on Stokes shift by 1.6 of the profile halfwidth. Moreover, the following values are taken for calculations in figures 2, 3: $(dn/dT)/C_p = -10^{-4}\,J^{-1}(cm)^3$; $t = 10^{-8}$ s; $n = 1.36$; $P_{32}/P_{21} = 100$; $\mu_{21} = 0.5$; $\mu_{32} = 0.000625$; $\lambda_{23} = 2\lambda_{12} = 1.0\mu m$; $\Delta\lambda_{23} = 4\Delta\lambda_{12} = 100nm$; ($\lambda_{ij}$; $\Delta\lambda_{ij}$ - are centre, halfwidth of the absorption band in i-j channel and $\omega_{ij} = 2\pi c/\lambda_{ij}$).

The pumping beam is taken with wavelength at the centre of absorption band for principal channel ($\lambda_0 = \lambda_{12} = 0.5\mu m$). Moreover, the radiations in the excited channel are taken with wavelength at the centre of excited absorption band ($\lambda = \lambda_{23} = 1.0\mu m$).

Figure 2 presents different components of phase response for dye solution as a function of the radiation intensity (I_{12}) in principal channel (S_0-S_1) at different intensities of radiations in excited channel (I_{23}). Figure 2 shows that, each component of phase response is saturated for pumping intensities more than saturation intensity in principal channel ($I_{12} > I_{12}^{sat}$). A hard

increasing of phase response happens, when the pumping intensity has a value near the

saturation intensity ($I_{12} \approx I_{12}^{sat}$).

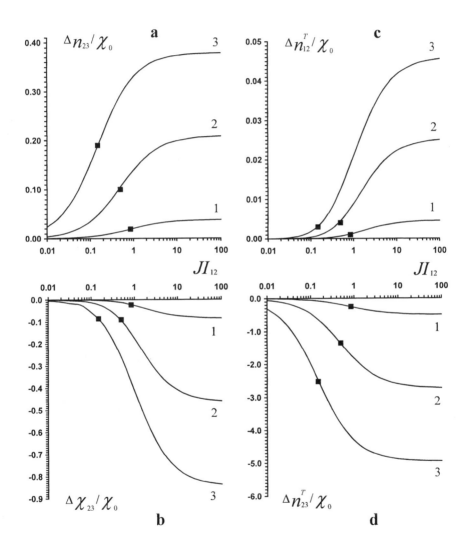

Figure 2 Dependence of resonance $\Delta n_{23}/\chi_0$ (a); $\Delta \chi_{23}/\chi_0$ (b) and thermal $\Delta n_{12}^{T}/\chi_0$ (c); $\Delta n_{23}^{T}/\chi_0$

(d) changes in refractive index on the radiation intensity in the principle channel JI_{12} for the radiation

intensity in the excited channel JI_{23}: 10; 100 and 1000 (carve: 1; 2 and 3 respectively).

166

Figure 3 presents changes in refractive index for dye solution as a function of the radiation intensity (I_{23}) in excited channel at different pumping intensity (I_{12}) in principal channel. In figure 3, each component of phase response reaches one half of its maximum value at

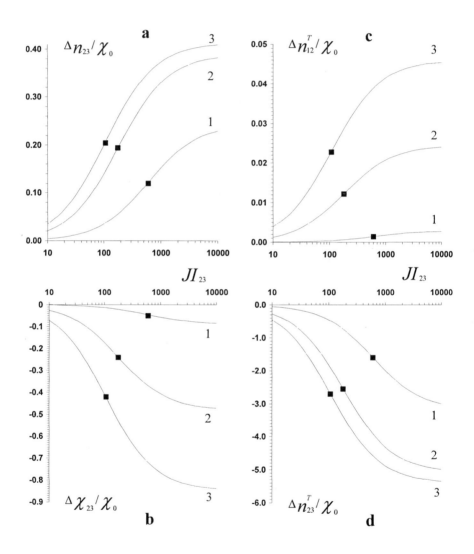

Figure 3 Dependence of resonance $\Delta n_{23}/\chi_0$ (a); $\Delta\chi_{23}/\chi_0$ (b) and thermal $\Delta n_{12}^T/\chi_0$ (c); $\Delta n_{23}^T/\chi_0$ (d) changes in refractive index on the radiation intensity in the excited channel JI_{23} for the radiation intensity in the principle channel JI_{12}: 0.1; 1.0 and 10.0 (carve: 1; 2 and 3 respectively).

saturation intensity in excited channel ($I_{23} = I_{23}^{sat}$). At the same time, the saturation intensity in excited channel is decreasing with increasing of intensity in principal channel, which agrees with equations (5, 7 and 8). From figures 2, 3 results that, the bleaching process in principal (excited) channel helps to reach the saturation in excited (principal) channel under law intensity in excited (principal) channel.

4. Conclusion

The phase response due to transitions between excited energy levels depends on intensity and frequency of radiations in both principal and excited channels (S_0-S_1 and S_1-S_2). Moreover, the bleaching process in one channel helps to reach the saturation in other one under law intensity in last channel. The saturation intensity in excited (principal) channel is decreasing with increasing of intensity in principal (excited) channel.

References

[1] Ivakin E V, Petrovich I N and Rubanov A S 1973 Russian J of Appl. Spectroscopy **18(6)**1003

[2] Agishev I N, Ivanova N A and Tolstik A L 1998 Optics Communications **156** 199-209.

[3] Haelterman M, Mandel P, Danckaert J, Thienpont I and Veretennicoff I 1989 Optics Communications **74** 238

[4] Karpuk S M, Romanov O G, and Tolstik A L 1999 Nonlinear Phenomena in Complex Systems **2(1)**50-55

[5] Kabanov V V and Rubanov A S 1981 Russian J of Appl. Spectroscopy **34(6)**975

[6] Pashinin P P, Sidorin V S, Tumorin V V and Shklovskill E I 1997 Quantum Electr. **27(1)**52-53

[7] Kabanov V V, Rubanov A S, Tolstik A L and Chaley A V 1986 Preprint Institute of Phys. of the Byelorutiassian Ac. of Sc.**411** 35

[8] Ben V N, Ivakin E V and Rubanov A S 1986 Preprint of the Institute of Phys. of the Byelorutiassian Ac. of Sc.**422** 8

Inst. Phys. Conf. Ser. No. 177
Paper presented at 1st Int. Conf. on Optical & Laser Diagnostics, London, 16–20 Dec. 2002

Saturation of Dye Solution in Principal Channel by Using Two Light Fields

Jihad S M Addasi

Department of Basic Sciences, Tafila Applied University College

Al–Balqa' Applied University

P.O. Box 179, Tafila-66110, Jordan

Abstract. A cubic nonlinearity of dye solution can be taken into consideration to study many nonlinear phenomena, such as phase conjugation, holography, amplification [1, 2, 5, 6 and 7]. Nonlinear processes strongly depend on intensity and frequency of radiations involved in these processes.

A dye solution can be modeled by a typical three-level configuration (S0-S1-S2) [2], for which the transitions of molecules in principal channel (S0-S1) are realized by the light fields of intensity I12 at frequency $\omega 0$. At the same time, light fields with intensity I23 at frequency ω interact with excited molecules to realize their transitions in excited channel (S1-S2). For this model, the conditions of phase response saturation are determined in principal channel.

The saturation of phase response of dye solution in principal channel (S0-S1) is realized at saturation intensity I_{12}^{saT}, decreasing with increasing of radiation intensity in excited channel I_{23}. The saturation intensity I_{12}^{saT} has its optimum (minimum) values when the frequency of light fields in principal channel is tuned with Stokes shift from the centre of principal absorption band. In addition, the saturation intensity I_{12}^{saT} has its optimum when the radiations in excited channel have enough big intensity I23 and a frequency tuning with anti-Stokes shift from centre of absorption excited band.

1. Introduction

A cubic nonlinearity of dye solution represents a basic information to study nonlinear processes such as: four-wave mixing, amplification and holography [1, 2, 3 and 5]. The lifetimes of vibration energy levels of dye solutions are significantly lower than the lifetimes of electronic energy levels [7]. In this case, the electronic states can be taken as homogeneously broadened levels, which gives ability to use three-level model with averaged Einstein coefficients for many nonlinear medium [2]. In three-level configuration the dye solution can be excited by light fields with two different frequencies, one group of light fields acts in principal (S_0-S_1) channel (with intensity I_{12} at frequency ω_0) and other group acts in excited (S_1-S_2) channel (with intensity I_{23} at frequency ω). Light fields in one channel can involve nonlinear processes, and other light fields act in second channel to control the nonlinear processes in first channel [3].

The balance equations under a double frequencies excitation of dye solution modeled by three-level configuration can be written as follows:

$$N_1 B_{12}(\omega_0) I_{12} = N_2 (B_{21}(\omega_0) I_{12} + v p_{21}) \; ;$$

$$N_2 B_{23}(\omega) I_{23} = N_3 (B_{32}(\omega) I_{23} + v p_{32}) ; \qquad (1)$$

$$N = N_1 + N_2 + N_3 \; ;$$

where

N_i – is the population of i- energy level; N – is the number of molecules in the unit volume of nonlinear medium; P_{ij} - is the total probability of spontaneous and radiationless transitions in the i-j channel; $v = c/n$ - is the light velocity in the nonlinear medium. The Einstein coefficients $B_{12}(\omega_0)$; $B_{21}(\omega_0)$ - are determined at the frequency of radiations ω_0 in principal (S_0-S_1) channel. At the same time $B_{23}(\omega)$; $B_{32}(\omega)$ - are determined at frequency of radiations ω in excited channel. The refractive index determined by balance equations can be used to study many nonlinear processes such as bleaching processes.

2. Theory

The saturation intensity I_{12}^{saT} in principal channel is defined as the value of radiation intensity, acting in principal channel, for which the absorption is decreasing in half of its initial value ($K_{12}(I_{12}^{sat}) = (1/2)K_{12}(I_{12} = 0)$). The extinction coefficient in principal channel at frequency ω_0 can be found by the following expression:

$$\chi_{12}(\omega_0) = \frac{c}{2v}K_{12}(\omega_0);$$ (2)

where

$K_{12}(\omega_0) = \frac{\hbar\omega_0}{v}(N_1 B_{12}(\omega_0) - N_2 B_{21}(\omega_0))$ - is the absorption coefficient in principal channel.

From equations (1) and (2) the extinction coefficient $\chi_{12}(\omega_0)$ will be:

$$\chi_{12}(\omega_0) = \chi_0(1 + aI_{23})/K;$$ (3)

where

$K = 1 + JI_{12} + aI_{23} + bI_{12}I_{23}$; $J = (B_{12} + B_{21})/vp_{21}$; $a = B_{32}/vp_{32}$; $b = B_{12}B_{23} + aJ$; $\chi_0 = N\hbar c B_{12}(\omega_0)/2v$;

$\chi_0(\omega_0)$ - is the linear extinction coefficient. The extinction coefficient, equation (3), has a monotonic proportionality with intensity of radiations in each channel (I_{12} and I_{23}) and has its maximum value ($\chi_{12} = \chi_0$) at intensities $I_{12} = I_{23} = 0$. The extinction coefficient has the half of its maximum value at saturation intensity in principal channel (I_{12}^{saT}) with value:

$$I_{12}^{saT} = \frac{1 + aI_{23}}{J + bI_{23}};$$ (4)

From equation (4) the saturation intensity (I_{12}^{saT}) has a monotonic dependence on radiation intensity in excited channel (I_{23}). To study the dependence of saturation intensity I_{12}^{saT} on frequency tuning of radiations in principal and excited channels, some parameters of nonlinear medium must be determined, especially this dependence has an optimum values as a function of frequency tuning.

3. Analysis and Discussion

Taking into consideration a nonlinear medium with a Gaussian form of mirror-symmetric absorption and emission bands on Stokes shift by δ_{ij} of the profile halfwidth Δ_{ij}. Where $\delta_{ij} = (\omega_{ij} - \omega_{ji})/\Delta_{ij}$, ω_{ij} - is the centre of i-j band. For this matter the frequency tuning of radiations in principal ($\eta_{12} = (\omega_0 - \omega_{12})/\Delta_{12}$) and excited ($\eta_{23} = (\omega - \omega_{23})/\Delta_{23}$) channels are used to find Einstein coefficients B_{ij}. Saturation intensity (I_{12}^{sat}) in principal channel for this form of absorption and emission bands has optimum values at frequency tuning of radiations, in principal (η_{12}^0) and excited (η_{23}^0) channels, with values:

$$\eta_{12}^0 = -\delta_{12}(B_{21}/vp_{21})\frac{1+al_{23}}{J+bl_{23}} ; \tag{5}$$

$$\eta_{23}^0 = \delta_{23}(al_{23}) .$$

All equations (1), (2), (3), (4) and (5) will be true for two-level model, when there is no radiations in excited channel ($I_{23} = 0 \Rightarrow N_3 = 0$). In this case let us rewrite equations (4) and (5) for two-level configuration:

$$I_{12}^{sat} = (1/J); \tag{4*}$$

$$\eta_{12}^0 = -\delta_{12}(B_{21}/vp_{21})/J \tag{5*}$$

In figures 1 and 2 are considered the following statements: the absorption and emission bands are on Stokes shift with values $(\omega_{ij} - \omega_{ji}) = \delta_{ij}\Delta_{ij} = 1.6\Delta_{ij}$, the maximum values of Einstein coefficients are the same for all bands $B_{12}^{max} = B_{21}^{max} = B_{23}^{max} = B_{32}^{max}$. The radiations intensities are normalized to the value vp_{21}/B_{12}^{max} and vp_{32}/B_{32}^{max} in principal and excited channels respectively ($I_{12n}^{(sat)} = I_{12}^{(sat)}B_{12}^{max}/vp_{21}$ and $I_{23}^n = I_{23}B_{32}^{max}/vp_{32}$). Figure 1 (a) and (b) show the monotonic dependence of saturation intensity I_{12}^{sat} on radiation intensity in excited channel I_{23}. From equations (4), (5), (4*) and (5*) the optimum conditions for saturation intensity are realized for: frequency tuning of radiations in principal channel in frequency diapason ($-0.21 < \eta_{12}^0 < 0$), and in excited channel more than zero ($\eta_{23}^0 > 0$), see figure 1 and 2.

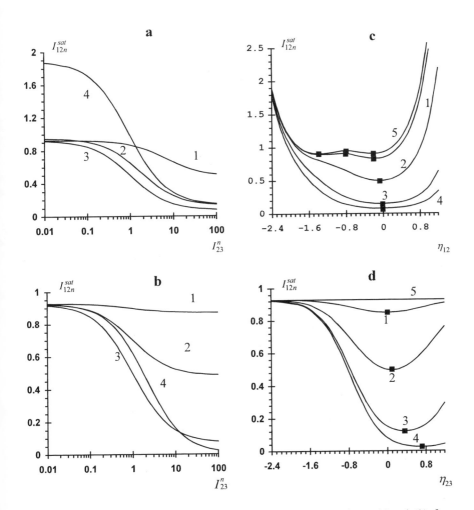

Figure 1- Dependence of saturation intensity I_{12n}^{sat} on: radiation intensity I_{23}^n (a) and (b); frequency tuning of radiation in principal η_{12} (c); in excited η_{23} (d) channels. Curves: 1, 2, 3 and 4 at: (a) η_{12}: -1.6, -0.8, 0 and 0.8; (b) η_{23}: -1.6, -0.8, 0 and 0.8; (c) and (d) I_{23}^n: 0.1, 1, 10, 100 respectively. Where for: (b), (d) η_{12}=0; (a), (c) η_{23}=0. Curve 5 is taken for two-level model.

Equation (4*) for two-level model is demonstrated by curve 5, and the extremum values of saturation intensity I_{12}^{sat} are determined by shaded squares in figure 1 (c) and (d).

Figure 2 represents the frequency tuning of radiations, for which are realized the extremum of saturation intensity, in principal channel η_{12}^0 (a), (b) and in excited channel η_{23}^0 (c) as a function of radiation intensity in excited channel I_{23}. To have a physical solution for the system of equations (5), the radiation intensity in excited channel must be positive ($I_{23} > 0$), that gives the following inequalities for frequency tuning: in principal channel

for minimum,

$$\frac{-\delta_{12}B_{21}}{B_{12}+B_{21}} < \eta_{12}^0 < \frac{-\delta_{12}B_{21}}{B_{12}(1+B_{23}/B_{32})+B_{21}}; \tag{6}$$

for maximum,

$$\frac{-\delta_{12}B_{21}}{B_{12}+B_{21}} > \eta_{12}^0 > \frac{-\delta_{12}B_{21}}{B_{12}(1+B_{23}/B_{32})+B_{21}}; \tag{7}$$

and in excited channel for minimum,

$$\eta_{23}^0 > 0 \tag{8}$$

Equation (5*) gives the values of frequency tuning η_{12}^0 for which are realized the extremum of saturation intensity I_{12}^{sat} for two-level model. For the three-level model the saturation intensity I_{12}^{sat}, as a function of tuning frequency η_{12}, has three extremums (two minimums and one maximum). For each of these extremums, the frequency tuning η_{12}^0 has upper and lower limits, which can be defined by inequalities (6) and (7). One of these limits $\eta_{12}^0 = -\delta_{12}(B_{21}/vp_{21})/J$ is realized at radiation intensity $I_{23} = 0$ and other limit

$\eta_{12}^0 \rightarrow \dfrac{-\delta_{12}B_{21}}{B_{12}(1+B_{23}/B_{32})+B_{21}}$ at $I_{23} \gg 1$. Figure 2 illustrates the dependence of frequency tuning

η_{12}^0 on radiation intensity in excited channel I_{23} for: (a) maximum in frequency diapason $-1.15 < \eta_{12}^0 < -0.8$ and local minimum in $-1.39 < \eta_{12}^0 < -1.15$; (b) optimum in frequency diapason $-0.21 < \eta_{12}^0 < 0$. Figure 2 (b) shows that, the increasing of radiation intensity in excited channel I_{23} gives a little anti-Stokes shift of optimum as a function of η_{12}. The same anti-Stokes shift happened for local minimum ($-1.39 < \eta_{12}^0 < -1.15$), but a Stokes shift happened for maximum with increasing of radiation intensity I_{23}. Figure 2 (a) shows that, for some frequency tuning η_{23} in excited channel curves 2, 3, 4, the local minimum and the maximum move closer one to other at frequency tuning $\eta_{12}^0 = -1.15$ (dashed line). In this case, both of the maximum and the local minimum are disappeared (figure 1 (c) curves 2, 3, 4 and figure 2 (a) curve 3).

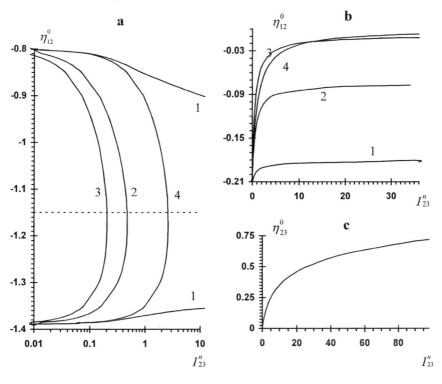

Figure 2- Dependence of frequency tuning: (a) and (b) η_{12}^0 in principal; (c) η_{23}^0 in excited channels on radiation intensity in excited channel I_{23}^n. Where curves: 1, 2, 3 and 4 are taken for frequency tuning of radiations in excited channel η_{23} : -1.6, -0.8, 0 and 0.8 respectively.

The upper part of figure 2 (a), above dashed line, corresponds to maximum and the lower part, under dashed line, corresponds to a local minimum.

The values of frequency tuning η_{23}^0 of radiations in excited channel, for which are realized the optimum values of saturation intensity in principal channel I_{12}^{sat}, are illustrated in figure 2 (c) as a function of radiation intensity I_{23}. Figure 2 (c) shows the monotonic dependence of frequency tuning η_{23}^0 on the radiation intensity I_{23}, where the anti-Stokes shift happens for optimum with increasing of radiation intensity in excited channel.

4. Conclusion

For double frequencies excitation of nonlinear medium, the saturation intensity in principal channel is decreasing with increasing of radiation intensity in excited channel I_{12}^{sat}. The optimization of saturation intensity is realized with the following statements:

1. The radiation intensity in excited channel must be enough big ($I_{23} > vp_{32}/B_{32}$).

2. The radiations in the principal channel must be tuned with a little Stokes shift from the centre of absorption band, not more than (1/5) of halfwidth of absorption band.

3. The radiations in excited channel must be tuned with a little anti-Stokes shift from the centre of absorption band, not more than halfwidth of absorption band.

References

[1] Ivakin E V, Petrovich I N and Rubanov A S 1973 Russian J. of Appl. Spectroscopy **18** N. **6** 1003.

[2] Agishev I N, Ivanova N A and Tolstik A L 1998 Optics Communications **156** 199-209.

[3] Addasi J S, Tolstik A L and Chaley A V 1993 Trans. of Byelorussion Univ. I N. **1** 9-12.

[4] Vasileva M A, Vishcakas J and Gulbinas V 1987 Quantum Electronics **14** 1691.

[5] Gancherenok I I, Romanov O G, Tolstik A L, Fleck B and Wenke L 2001 J. of Opt. B. **3** 220-224.

[6] Rubanov A S, Tolstik A L, Karpuk S M and Ormachea O 2000 Optics Commun. **181** 183-190.

[7] Tichonov E A and Shpak M T 1979 Nonlinear Optical Effects in Organic Compounds Kiev Naukowa Dumka 90-100.

Inst. Phys. Conf. Ser. No. 177
Paper presented at 1st Int. Conf. on Optical & Laser Diagnostics, London, 16–20 Dec. 2002

Particle image velocimetry: from two-dimensional to three-dimensional velocity mapping with holographic recording, object conjugate reconstruction (OCR) and optical correlation

N A Halliwell

Wolfson School of Mechanical and Manufacturing Engineering, Loughborough University, UK

Abstract. Particle Image Velocimetry (PIV) techniques have been developed to provide routine two-dimensional velocity mapping in hostile flow situations. Examples of technique developments and results obtained from in-cylinder internal combustion engine flows will be presented. Holographic Particle Image Velocimetry (HPIV) in conjunction with complex correlation processing will be introduced. This is a means of obtaining high resolution, three dimensional velocity vector maps throughout a flow volume whilst avoiding problems caused by image distortion. The approach effectively combines traditional Holographic Interferometry and Velocimetry and, for the first time, allows for simultaneous velocity measurements in both fluid and solid mechanics. Recent research work has provided a new technique for the recovery of data from holographic recordings through Object Conjugate Reconstruction (OCR). This will be introduced and shown to offer significant advantages in automating data extraction from a flow volume.

Inst. Phys. Conf. Ser. No. 177
Paper presented at 1st Int. Conf. on Optical & Laser Diagnostics, London, 16–20 Dec. 2002
©*2003 IOP Publishing Ltd*

Imaging Fibre Bundle Based 3 Component Planar Doppler Velocimetry

David S Nobes and Ralph P Tatam

Optical Sensors Group
Centre for Photonics and Optical Engineering
School of Engineering, Cranfield University
Cranfield MK43 0AL UK.

Abstract. A planar Doppler velocimetry (PDV) technique is described that is capable of measuring the three, instantaneous components of velocity in a plane using a single pair of signal and reference cameras. PDV can be used to measure the instantaneous 3-D velocity of a fluid by using a frequency-to-intensity converter to determine the Doppler shifted frequency of the output of a narrow line-width, pulsed laser that has been scattered off particles seeded into the flow. In the technique presented here the three views required to obtain three dimensional velocity information are ported from the collection optics to a single image plane using flexible coherent imaging fibre bundles. A fourth channel of the fibre imaging bundle is used to image individual laser pulses and allow correction for pulse-to-pulse frequency variations. To demonstrate the technique results are presents of the instantaneous velocity field of a jet.

1. Introduction

Advances in flow measurement diagnostics has allowed fluid mechanics researchers to come closer to the ultimate experimental scenario of fully resolved flow information over a volume of the flow as the flow changes in time. While this has yet to be achieved, advances in measurement hardware have allowed the development of techniques that capture a snapshot of flow properties over a two-dimensional area. One of these, planar Doppler velocimetry (PDV)[1-4] is a velocity measurement technique that is capable of measuring the three instantaneous components of velocity over a two dimensional plane.

The theory of PDV is based on the Doppler principle that light, when scattered of a moving object is frequency shifted depending on the velocity of that object. The Doppler formula describes the relationship between the velocity of the object and the frequency shift of the light scattered from that object and can be expressed as, $\Delta \upsilon = \frac{\upsilon_0 (o - i) \bullet V}{c}$. Here, $\Delta \upsilon = \upsilon_D - \upsilon_0$ is the difference between the Doppler shifted frequency (υ_D) and the frequency of the light source (υ_0), c is the free space speed of light, V is the velocity vector of the object that is scattering the light, o is a unit direction vector to the viewer and i , the unit direction vector of the light source.

From the Doppler formula it can be seen that to determine a single component of velocity of the scattering object, the Doppler shift of the scattered laser needs to be determined as well as having knowledge of the propagation direction of the light source and the direction from which the scattered signal is collected. To determine the three components of velocity at a point, scattered light from the same single laser source can be viewed from three different directions each measuring a different component of velocity. If the laser source is spread into a two-dimensional sheet and a region of interest is viewed from three different directions using two-dimensional array detectors such as CCD cameras, then the three components of velocity can be determined over a plane. PDV does not have to clearly image the particles seeded into a flow used as the scattering medium as with the other major planar velocity technique, particle imaging velocimetry (PIV)[1]. It is only the frequency of the scattered light that needs to be measured, resulting in less stringent requirements for the imaging system.

While the light source and viewing direction can be readily determined a method for determining the Doppler shift of the light frequency is needed. The PDV technique measures the shift in frequency directly using a frequency-to-intensity converter. Essentially this is a filter that attenuates the scattered signal based on the frequency of that signal. For this filter to be efficient, it must have a cut-off that extends over the maximum expected range of velocities. For this a basic example of the expected performance of the filter can be developed. If a laser at a wavelength of 532nm is used as the light source and the minimum to maximum expected velocity range is +/- 100m/s then the Doppler shift will be +/- 200MHz. In this example the filter will then need to have maximum cut-off for just over 400MHz. For such a narrow operation range Komine[6] identified the use of the atomic and molecular absorption bands and identified iodine vapour as the most appropriate for operation with Argon and frequency doubled Nd:YAG lasers. The philosophy of the PDV technique is to tune a narrow line width laser light source to a point that is half way either up or down the side of an absorption band[1-4]. For no shift in the laser frequency, the detector will see a signal of 50% the value of full transmission. Any Doppler shift in the laser frequency will be measured as a change in the signal intensity level. The sign of the direction can also be determined from the direction of the shift to either lower or higher intensities.

Early in the development of PDV it was realised that a reference was needed to normalize the signal attenuated by the filter to account for variations in the scattered intensity[6]. Initial work used CW lasers[7,8], however many research groups quickly moved to pulsed, frequency doubled Nd:YAG systems with the main aim of increasing the available laser power while at the same time making instantaneous measurement possible. Meyers and Komine[7] began the initial work in the area using a CW Argon-ion laser and their technique is perhaps the classic arrangement of PDV. Here, six cameras in three pairs of signal and reference cameras viewed a region of interest from three different directions, each pair utilizing its own iodine cell. Frequency drift of the laser is an important problem, especially for the pulsed Nd:YAG systems[9,10]. For these lasers an injection seeder laser is used to ensure single mode operation and tune the host laser. To help lock the host laser to the seed laser the length of the cavity of the host laser is adjusted. This is done in a feedback manner with the position of one of the mirrors of the host being dithered. The resultant is an output that can be tuned over the gain curve of the host laser by the seed laser and dithers about the frequency of the seed laser. The amount of dither is dependent on the set-up of the whole laser system but can be of the order of 100MHz. To counter this movement of the host laser frequency many research groups have employed a system to measure the frequency of the laser for the pulse that is used to take the measurement. A typical system is an arrangement of single point detectors and a separate iodine cell[2,4].

2. Experimental Set-up

A schematic of the experimental arrangement used in this work is shown in Figure 1[11]. A pulsed, injection seeded, frequency doubled Nd:YAG laser is used as the light source and allows for the capture of instantaneous velocity measurements. Each laser pulse is spread into a light sheet using a combination of spherical and cylindrical lenses. A unique feature of this PDV system is that a single head is utilized consisting of two cooled CCD cameras and a single iodine cell in a heated oven to obtain three velocity components and the laser pulse frequency. The iodine concentration in the cell is controlled by temperature control of the 'cold', side finger of the cell. The imaging optics shown in Figure 1 port an image into the head system that is first split into two paths to the signal and reference CCD's with a 50/50 non-polarising beam splitter. The image ported through the detector head is delivered to it by an imaging fibre bundle. This is a coherent array of fibres that is split into four channels. Each channel has 600x500 fibres that are 8μm in diameter and at 10μm centres. Three of these are used to view the same region of interest from three different directions providing the components of velocity. The fourth as shown in Figure 1, images the small percentage of the laser pulse that has passed through a 532nm dichroic mirror and scattered off a screen. This provides a means of monitoring the frequency of each laser pulse. To demonstrate the combining of images an example image of the combined three views and the laser beam reference view is shown in Figure 2. The top left (View A) and bottom right (View B) and left (View C) are the three views of a reference target of crosses and views an area ~150x150mm. The perspective distortion of the target highlights the different directions of the three views. The top right corner of the image is the fourth channel used to image the laser beam. For this configuration of equipment each of the three views from the imaging fibre bundle has slightly more pixels per view than the six-camera system described earlier.

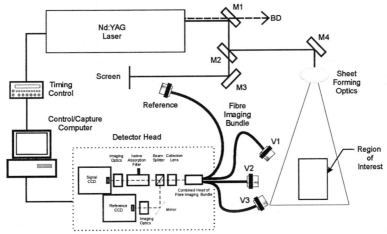

Figure 1. A schematic of the experimental arrangement for the PDV experiment. ___ 532nm beam; -- 1064nm beam; M1&M2 – 532nm dicroic mirrors; M3&M4 – turning mirrors, BD – 1064nm beam dump

184

3. Results

Instantaneous, three component velocity measurements of a jet flow have been collected using the current PDV system. The jet was oriented vertically up and had an exit diameter of 70mm with a nominal exit velocity of 30m/s resulting in a Reynolds number of $> 1.4 \times 10^5$. The jet was seeded with smoke particles with a nominal mean diameter of $5\mu m^5$.

Figures 3&4 show the raw images collected by the signal and reference cameras respectively. The signal image in Figure 3 shows a general attenuation compared to Figure 4 due to the laser being tuned to 50% of the full maximum of the Iodine absorption spectrum. Overlaid with this is attenuation due to the Doppler shift of the laser frequency. In the top right corner of each image is an image of the laser pulse, which is used to determine the frequency of the laser. View A and B have been viewed from the same side of the laser sheet while View C was located on the other side of the sheet. A result of this can be seen in Figures 3&4 where in views A&B the jet is leaning to the right while in View C the jet is leaning to the left.

To process velocity measurements each of the raw images are filtered with a 3x3 smoothing filter and then each view is cut out of the images and handled separately to be de-warped and mapped onto a common scale. The three views are then divided and corrected for variation in the laser power across the laser sheet and for the response of the camera over it's A/D range. The results are converted from an intensity measurement to frequency by using a polynomial function to describe the transfer function of the Iodine absorption spectrum. This frequency and the length of the sensitivity vector (*o-i*) are then used in the Doppler formula to calculate the velocity measurement captured by each view. The data from the three non-orthogonal sensitivity vectors is then mapped to an orthogonal co-ordinate system using a transformation matrix[3].

 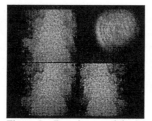 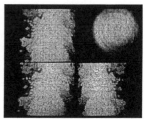

Figure 2. An example image of the view through the image bundle of a reference target.

Figure 3. Raw image of the jet flow from the signal camera.

Figure 4. Raw image of the jet flow from the reference camera.

The resultant instantaneous velocity field for a region inside the jet is shown in Figure 5. In this image only every 5th computed vector has been shown. There is a general trend of the vectors in the field pointing in the main flow direction of the jet with a comparable magnitude with that expected to be measured in this flow. The turbulent nature of the flow is evident by the varying magnitude and direction of the velocity vectors defining the in-plane velocity components. The out-of-plane Z-component of velocity defined by the colour contour plot in the background shows that the flow while globally has no main velocity component that there are local regions of both in and out-of-plane motion. A close-up of a region defined in Figure 5 is shown in Figure 6 where all computed vectors are displayed. Figure 6 shows that the computed vectors at this scale have a smooth transition in magnitude and direction across the flow. In the Figure a local reverse flow is present with an associated vortex.

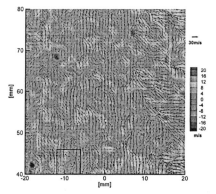

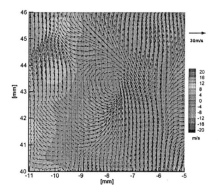

Figure 5. An instantaneous three component velocity field - only every 5th vector shown. In-plane components show as vector and out-of plane show in the colour map.

Figure 6. A close-up of a region in the flow (defined in Figure 5) with every computed vector shown.

4. Conclusions

A PDV technique has been described that is capable of measuring the three instantaneous components of velocity in a plane. All three views needed to define a three-dimensional velocity field have been recorded onto a single pair of signal and reference cameras by porting the three views through an imaging fibre bundle. The legs of this bundle combined into a single array that is imaged through the PDV detection head. Laser frequency stability is monitored and measured using the fourth leg of the bundle allowing for correction for any frequency movement of the laser for each instantaneous measurement. The use of a single PDV detector head minimises the number of CCD array cameras that are used to two and only a single iodine cell is used for both measurement the three velocity components and referencing measurement the laser frequency.

Acknowledgments

The work was funded by the Engineering and Physical Sciences Research Council (EPSRC), UK, and by the Royal Society.

References

[1] Samimy, M. and Wernet, M.P. AIAAJ, Vol. 38, No4, pp 553-574, 2000.
[2] Mosedale, A.D., Elliott, G.S., Carter, C.D. and Beutner, T.J. AIAAJ, Vol 38, No6 pp 1010-1024, 2000.
[3] Reinath, M.S., NASA TM-112210, 1997.
[4] Elliot, G.S. and Beutner, T.J. Prog in Aero. Sci. Vol 35 pp 799-845, 1999.
[5] Meyers, J.F., von Karman Institute Lecture Series 1991-08, 1991.
[6] Komine,H. US Pat No. 4,919,536, 1990.
[7] Meyers, J.F. and Komine, H. Laser Anemometry, vol. 1, pp. 289-296, 1991.
[8] Ford, H.D. and Tatam, R.P. Optics and Lasers in Engineering, Vol 27, pp. 675-696, 1997.
[9] McKenzie, R. L., Applied Optics, Vol. 35, No. 6, pp. 948–964, 1996.
[10] Forkey, J.N., Lempert, W.R. and Miles, R.B. Opt Lett, Vol 22, pp. 230-232, 2001
[11] Nobes, D.S., Ford, H.D. and Tatam, R.P. (2002) 40th AIAA, Reno, Nevada, 14-17 Jan 2002

Inst. Phys. Conf. Ser. No. 177
Paper presented at 1st Int. Conf. on Optical & Laser Diagnostics, London, 16–20 Dec. 2002
©2003 IOP Publishing Ltd

Optical-Diagnostic Study of Two-Dimensional, Premixed Laminar Flames

M Fairweather and [†]G K Hargrave

Department of Chemical Engineering, University of Leeds, Leeds LS2 9JT, UK
[†]Wolfson School of Mechanical and Manufacturing Engineering, Loughborough University, Loughborough LE11 3TU, UK

Abstract. Gas burning appliances used in the domestic and commercial markets incorporate laminar flame burners that are made up of a number of ports which give rise to multiple laminar flames. Virtually all current burners use premixed natural gas-air mixtures giving rise to partially or fully aerated flames. The increasing demand for improvements in appliance efficiency and reduction in emissions has meant that combustion on such burners has been of renewed interest lately. Advanced laser diagnostic and computational modelling techniques are now being used to develop a basic understanding of how burner design affects combustion performance. This paper presents the results of a detailed optical-diagnostic study of the structure of two-dimensional, premixed laminar flames stabilized on a multi-slot burner. Measurements have been made in fuel-lean, stoichiometric and fuel-rich flames, with detailed data obtained using Raman scattering for major chemical species, laser-induced fluorescence for minor species, and coherent anti-Stokes Raman spectroscopy for temperature. The data gathered give insights into the structure of these practically relevant flames, and are also of value in the validation of computational models for use in burner design.

1. Introduction

The vast majority of natural gas burning appliances used in the domestic and commercial markets incorporate laminar flame burners. Conventionally, these burners are made up of a number of ports, in the form of slots or holes, which give rise to multiple laminar flames which may or may not interact depending upon the particular burner design. In addition, because of the flame stability and gas interchangeability problems encountered when using non-premixed flames, virtually all current burners use premixed natural-gas air mixtures that give rise to partially or fully aerated flames.

The increasing demand for improvements in appliance efficiency, reduction in emissions and more compact appliance designs has meant that combustion on laminar flame burners has been of renewed interest lately. In particular, the availability of advanced laser diagnostic and computational modelling techniques has provided the opportunity to develop a basic understanding of how burner design affects combustion performance, thereby facilitating the design and development process. This is in contrast to the conventional, and time consuming, empirical approach to burner design in which combinations of parameters, such as port geometry, port loading and primary aeration, are varied, and an optimum design inferred from observations of flame stability and emission measurements.

Computational models, in particular, are now capable of providing detailed simulations of the characteristics of laminar flames, and as a consequence are being used increasingly in the design of laminar flame burners. Recent years have seen the development of a number of models, e.g. [1-3], that are capable of providing detailed predictions within, and in close proximity to, the reaction zone of multi-dimensional laminar flames. In order to ensure the reliability and accuracy of such models, however, it is necessary to test their predictions against experimental data that is of the same level of detail as that required for design purposes. In doing this, detailed experimental information is required not only on idealized flames, e.g. [4, 5], but also on flames stabilized on burners of realistic design.

The present paper describes a detailed optical-diagnostic study of the structure of two-dimensional, premixed laminar flames stabilized on a multi-slot burner that provides a realistic representation of burners used in actual appliances. This burner has been employed previously in both experimental [6] and computational [3] studies, with earlier work [6] having provided measurements of the velocity field within various stoichiometry flames, and of temperature and CO and NO levels along their centre-line. This study complements and extends earlier work by providing detailed measurements of temperature and major and minor species in flames located on the central slot of the multi-slot burner.

2. Experimental work

The burner used as the basis of the study was a five-slot version of a single-slot design used by Beyer and DeWilde [7] that generated a translationally symmetric, two-dimensional flame on the central slot. Depending on the primary aeration used, this central slot flame is similar to those found on fully premixed burners, or to the inner cones of practical partially premixed burners. The burner was made entirely of brass, and consisted of a plenum, flame trap and five 3mm wide×50mm long slots, each separated by a 3mm thick plate. The flame trap also served to distribute the inlet flow uniformly, with each slot containing flow conditioning fins at its base. Like the original design, the burner employed ports at the end of each slot that allowed an inert shielding flow of nitrogen to be used to eliminate secondary air entrainment. The complete design was not only capable of yielding stable flames for primary aerations ranging from fuel-lean to fuel-rich conditions, but also resulted in an innermost flame that was effectively shielded from the ambient environment. In the experiments performed, fully premixed mixtures of methane and air were delivered to the burner via rotameters, with a fixed flow rate of air having been used and the flow of methane adjusted to achieve the desired aeration. Three aerations were considered in the experiments: a fuel-lean flame (125% primary aeration (p.a.), equivalence ratio, $\phi = 0.8$), a stoichiometric flame (100 % p.a., $\phi = 1.0$), and a fuel-rich flame (83 % p.a., $\phi = 1.2$). Previous work [6] demonstrated that the flow exiting the various slots was uniform, with a parabolic velocity profile, and that the flame formed on the central slot was two-dimensional and laminar. Operation of the burner was found to result in significant increases in its temperature, and in the present work solid temperatures measured by thermocouples embedded in the plates that formed the central slot were allowed to achieve steady state values prior to commencing measurements. Solid, steady state temperatures of 395, 430 and 406K were recorded for the fuel-lean, stoichiometric and fuel-rich flames respectively.

Measurements were made mid-way along the central slot flame, with the burner mounted in a three-dimensional traverse that allowed the flame to be translated relative to the measurement systems. Data were obtained using non-intrusive, laser-based optical-diagnostic techniques, including Raman scattering, laser-induced fluorescence (LIF) and coherent anti-Stokes Raman spectroscopy (CARS).

Major species (CH_4, CO, CO_2 and O_2) concentration data were obtained using Raman scattering. The light source for the system was an injection-seeded, frequency-doubled Nd:YAG laser which provided 450mJ in a 10ns pulse at 532nm. The laser light was focused to the point of interest using a 1500mm spherical lens, generating a beam waist at the control volume of 700µm. Scattered light was collected using f3 optics and focused onto the entrance slits of a 0.5m spectrometer incorporating an 1800line mm^{-1} holographic grating. The spectrometer was fitted with a gated, peltier-cooled ICCD providing 580 pixels configured across the Raman spectra, and vertical binning of 384 pixels to improve signal to noise ratio. With this configuration it was not possible to capture all the required spectra in a single measurement, and the spectra were therefore scanned to capture 10 individual spectra to build the information for the species considered. At each measurement point the Raman light was averaged over 250 laser pulses. The intensity of the light scattered perpendicular to the input laser beam was obtained using the approach described by Ebersohl et al. [8], with combined efficiencies measured by calibration with pure gases and gas mixtures.

The LIF system was configured to provide one-dimensional measurements for CH and NO, and two-dimensional imaging for OH. The laser light was provided by an Nd:YAG laser, dye laser and doubler arrangement. For the CH measurements, the laser was operated frequency tripled to 355nm and pumped a coumerin 430 dye laser providing 20mJ $pulse^{-1}$. The NO measurements were obtained by frequency doubling the dye laser output to pump the 226nm transition. For both the CH and the NO fluorescence measurements the laser light was focused using an $f = 1500$mm spherical lens. A 10mm length of the beam was imaged onto a 580×384 pixel ICCD camera, and the signal binned over the 1mm width of the beam. The ICCD was gated to capture fluorescence over a 100ns period after laser excitation. The fluorescence signal was averaged over 100 individual laser pulses. The OH LIF data was obtained by two-dimensional imaging. The laser source was an Nd:YAG laser pumping a tunable dye laser operating with Rhodamine 620, frequency doubled to 308nm. The light was formed into a 20mm×1mm sheet and the fluorescence signal imaged over 15mm×10mm using a 580×384 pixel ICCD. The image was averaged over 100 laser shots.

The CARS system used was a broadband BOXCARS arrangement. The laser source was a Continuum Powerlite 8000 laser generating up to 800mJ at 532nm. The laser was injection seeded with a line width of 0.05cm^{-1} at 532nm, and was run with a maximum output pulse energy of 320mJ in a 10ns pulse. This beam was split, with 120mJ passing via a delay line to the pump beams and 200mJ pumping a Mode-X modeless broadband dye laser operated with Rhodamine 640 laser dye. The dye laser was operated with a four-pass oscillator configuration and a single amplifier. This generated a 45mJ $pulse^{-1}$ in the probe beam. All laser powers were monitored using a calibrated Ophir pulsed power meter. The broadband configuration probes all of the available nitrogen transition simultaneously by using the broadband probe laser beam from the dye laser. The collimated beam geometry passing to the final focusing lens consisted of the two pump beams and a single probe beam arranged on the vertices of an equilateral triangle of side 25mm. The final focusing lens was 50mm in diameter and 250mm focal length which generated a control volume in the form of a cylinder 0.8mm long and 0.4mm in diameter. The CARS signal beam was passed through a dichroic filter which reflected the 532nm pump and 640nm probe beams and transmitted

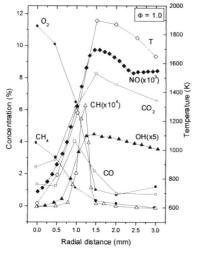
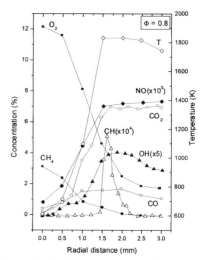

Figure 1. Radial temperatures and species concentrations in the fuel-lean flame.

Figure 2. Radial temperatures and species concentrations in the stoichiometric flame.

the blue CARS signal beam. The signal was passed directly to a SPEX 1.25m spectrometer. The spectrometer was fitted with a 2400grooves mm^{-1} grating which generated a specific dispersion of 0.05nm mm^{-1} at the spectrometer output. The detector for the spectrometer was a Princeton Instruments back-thinned CCD with 1750×550 pixels and a resolution of 16 bits. The detector was arranged in a horizontal configuration, with the 1750 pixels across the CARS spectrum. The detector had 13μm wide pixels, with the spectrometer/detector configuration providing a resolution of 0.003nm pixel^{-1}. To reduce thermal and electronic noise the detector was cooled to 263K. The CCD detector was synchronised to the laser source to provide a single CARS spectrum record for every laser pulse. The CARS spectra were recorded using the Princeton Instruments WinSpec software which can store spectra at a rate of 6Hz. At each measurement location 500 CARS spectra were recorded over approximately 90s. The Sandia National Laboratories fitting code FTCARS was used for comparison with the theoretical CARS spectra and for determination of flame temperature.

3. Results and discussion

Axial and radial profiles of temperature and chemical species were obtained in the three flames. Attention is focused on radial measurements made 2mm above the burner exit, roughly half-way along the length of the three flames, in order to elucidate their structure. Axial temperatures and NO levels, through the flames and into the post-flame region, are also considered due to the importance of such data to emissions considerations.

The degree of aeration of a flame critically affects its structure and combustion chemistry. Low aerations (fuel-rich) are representative of the majority of current gas burners, although highly aerated (fuel-lean) flames are becoming more common because of their potential for significantly reducing NO$_x$ emissions. These changes in flame characteristics are reflected in the results of Figures 1 to 3 that give, respectively, radial measurements obtained in the fuel-lean, stoichiometric and fuel-rich flames. In all three flames, CH$_4$ and O$_2$ levels decrease with increasing radial distance from the flame centre-

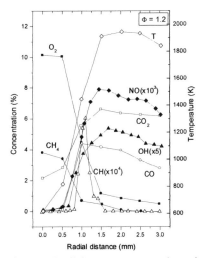

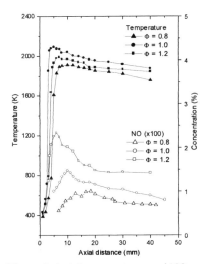

Figure 3. Radial temperatures and species
concentrations in the fuel-rich flame.

Figure 4. Axial temperatures and NO
profiles.

line. Within the cool, unburned regions of the flow, temperatures at first increase slowly
with radial distance, but then rapidly as the reaction zone is approached. Highest
temperatures are achieved in the fuel-rich flame, although this is due to the axial location
chosen for comparison purposes being at an effectively different position within each flame
when considered in terms of its structural development. Peak axial temperatures, considered
further below, are in line with anticipated adiabatic flame temperatures for the various
equivalence ratios considered. Temperature profiles indicate that the reaction zone thickness
is in agreement with burning velocity considerations, with the stoichiometric flame being
thinnest, and the fuel-lean flame thickest. This order in flame thickness is also reflected in
the CH radical results, with this species, together with flame temperature, being important in
the NO formation process since its reaction with N_2 is considered the rate controlling step in
the prompt NO pathway. Hydroxyl radicals, as a combustion intermediate, also indicate
chemical activity in the reaction zone. CO is formed as an intermediate in the combustion
process, and remains at significant levels after the flame front in the fuel-rich flame, whilst
in the stoichiometric case levels rapidly decay. For the fuel-lean flame, CO remains at a low
level over all radial positions examined. As a stable product, CO_2 reaches similar levels in
all the flames, although its distribution varies with aeration. NO levels are greatest in the
highest temperature stoichiometric flame, with the fuel-lean flame showing the lowest levels.
NO formation is intimately linked to temperature, and the radial NO profiles are seen to
follow qualitative temperature variations within the flames, with peak NO levels roughly
coinciding with temperature maxima.

The axial temperatures of Figure 4 demonstrate that, in line with burning velocity
considerations, the relatively thin reaction zone has its tip stabilized closest to the burner in
the stoichiometric case, with the fuel-lean flame tip being located furthest downstream.
Temperatures along the centre-line rise fairly rapidly to peak values, with the stoichiometric
and fuel-lean flames again exhibiting, respectively, the thinnest and thickest reaction zones.
Above the tip of the flames, temperatures decay steadily due to cooling by radiation heat
transfer. The initial rise in NO with increasing flame temperature is due to the prompt
formation mechanism and coincides with the formation of CH, with levels in all the flames

decreasing after the peak in axial temperatures. The Zeldovich thermal mechanism for NO formation becomes important at temperatures between 1800 and 2000K, with an excess of oxygen also being required. This is because the most important kinetic step in the formation process has a high activation energy, with oxygen concentrations also depending strongly on temperature. The present results indicate little or no growth in NO through the thermal mechanism in the post-flame region. This is to be anticipated in the rich and lean flames due, respectively, to the absence of oxygen and low post-flame temperatures. Increases in NO through the thermal mechanism have been observed in a near-stoichiometric flame [6] similar to that considered in the present work. Significant differences do, however, occur between temperatures measured in these flames, with a peak axial temperature of $\approx 2100K$ (which rapidly decays below 2000K) obtained in the present work contrasting with a value of $\approx 2300K$ in the earlier [6] study. Given these differences in measured temperatures, the lack of post-flame NO growth in the present study is not inconsistent with earlier results.

4. Conclusions

A detailed optical-diagnostic study has been performed on a number of different equivalence ratio, two-dimensional, premixed laminar flames stabilized on a multi-slot burner. Data has been gathered using Raman scattering for major chemical species, laser-induced fluorescence for minor species, and coherent anti-Stokes Raman spectroscopy for temperature. The data gathered gives insights into the structure of these practically relevant flames, and highlights the influence of aeration on flame structure and emission levels. The data is also of value in the validation of computational models for use in burner design.

5. Acknowledgement

The authors would like to thank Advantica Technology for providing the burner used in this study, and Dr. S.M. Hasko for his assistance and helpful discussions.

References

[1]   Katta V R and Roquemore W M 1995 *Combust. Flame* **102** 21-40
[2]   Bennett B A V and Smooke M D 1998 *Combust. Theory Modelling* **2** 221-258
[3]   Brown M J, Fairweather M and Hasko SM 1999 *J. Inst. Energy* **72** 89-98
[4]   Gore J P and Zhan N J 1996 *Combust. Flame* **105** 414-427
[5]   Nguyen Q V, Dibble R W, Carter C D, Fiechtner G J and Barlow R S 1996 *Combust. Flame* **105** 499-510
[6]   Hasko S M, Fairweather M, Bachman J-S, Imbach J, van der Meij C E, Mokhov R V, Jacobs R A A M and Levinsky H B 1996 *Proc. 1995 Int. Gas Res. Conf., Cannes vol 2* (Rockville: Government Institutes Inc) p 1390
[7]   Beyer R A and DeWilde M A 1982 *Rev. Sci. Instrum.* **53** 103-111.
[8]   Ebersohl N, Klos T H, Suntz R and Bockhorn H 1998 *Proc. Combust. Inst.* **27** 997-1005.

Inst. Phys. Conf. Ser. No. 177
Paper presented at 1st Int. Conf. on Optical & Laser Diagnostics, London, 16–20 Dec. 2002
©2003 IOP Publishing Ltd

In-Plan Quantification of 3-D Velocities and Turbulence Intensities using a Flow Visualisation Technique

A Ramadan[†], S Michelet*, K C Lee* & M Yianneskis*

[†] Corresponding author
Centre for Aeronautics, School of Engineering, City University London,
Northampton Square, London, EC1V 0HB

* Experimental and Computational Laboratory for the Analysis of Turbulence
(ECLAT), Department of Mechanical Engineering, King's College London,
Strand, London, WC2R 2LS

Introduction

Most engineering applications involve complex, three-dimensional flow patterns which require the employment of non-intrusive measurement methods if the mean flow and turbulence characteristics are to be accurately determined. One such example, which has provided the motivation for the present work, is the characterisation of the flows in stirred vessels. Here the rapid acquisition of data is essential for the practical assessment of the mixing capabilities of different stirrer configurations, especially where the presence of flow macro-instabilities necessitates the rapid acquisition of time-resolved flow patterns. As an example, the work conducted by [1] has indicated the presence of macro-instabilities in the flow produced by a hydrofoil impeller which can only be properly characterised and quantified with a suitable real-time, three-dimensional, non-intrusive method.

The 3-D flow visualisation technique reported in this paper is capable of providing rapidly quantitative as well as qualitative 3-D information, with relatively inexpensive and simple to install and operate equipment. The analysis algorithm employed by the technique relies on determining the length and colour of the resulting tri-colour streaklines to provide field-wide velocity information which is not subject to ambiguity stemming from varying particle concentration fields. The present system comprises a conventional CCD colour camera, a single laser sheet and two colour light sheets. In addition to the advantage of the simplicity of apparatus, the data retrieval is straightforward as no calibration is required. The technique has been successfully applied to measure the three mean velocity components (U, V and W) and the associated turbulence levels (u, v and w) and kinetic energy of turbulence K of an inclined 45° circular jet contained in a rectangular enclosure. The technique has also been further developed and fine-tuned to characterise the flow patterns in a cylindrical stirred vessel (although this experiment is not the subject of this communication). The principles and methodology of the technique are described in the following sections and characteristic results from the application to the above test case are presented and compared with similar data obtained with 3-D PIV. The capabilities and accuracy of the technique are assessed and possible future developments to improve its accuracy and resolution are identified.

Experimental configuration

The experiment used to develop and validate this technique made use of an inclined jet of water discharging into a rectangular enclosure from a 10mm diameter circular tube. The flow rate through the jet was 5.4 l/min. A schematic representation of the imaging system is shown in Fig. 1 below. A white light source (Thorn, 1000 W) with 2 colour filters was used to generate two light sheets (yellow and red, each of 3mm thickness). An Argon Ion laser (Coherent Innova 90 at a wavelength of 514.5 nm) was used to generate a laser sheet of average thickness (Δ) of 1.63mm, which was located between the yellow and red light sheets. A CCD colour camera (Type EVI Model 1011P) was used to record the images (at 768×576 pixels resolution) at a framing rate of 25 Hz. Polystyrene particles with diameters in the range 250μm-350μm were used as light scatterers. The inclined jet configuration and the plane in which visualisation images were obtained were selected so that the flow would have an out-of-plane velocity component comparable in magnitude to one of the in-plane components and there would be no recirculation of the flow within the plane of observation, so as to facilitate the development and validation of the technique.

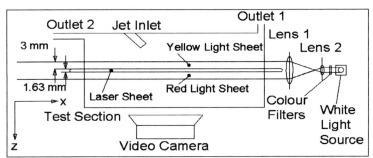

Figure 1 Schematic plan view of optical arrangement (not to scale)

The experiment was started by switching the pump on and adjusting the valves (at the jet inlet and the two outlets) until the required flowrate, which was monitored by a variable-area flowmeter, was achieved. The system was then left running for steady-flow conditions to be reached. The tri-colour streaklines across the measurement area were then directly recorded onto a video VHS tape. A total of 2147 images resulting from 270 seconds recording time were captured and digitised off-line using an Image-Grabber Card (Data Translation DT3153 controlled by Impuls-Vision XXL software) on a PC. The images were grabbed in the same sequence as the one on the VHS tape and saved in tiff format. Each image is 18mm in the x-direction and 14mm in the y-direction, hence each pixel represents an area of approximately 23μm by 24μm.

Principle of technique

The method relies on the determination of the length of the path travelled by a particle during the exposure time of the camera (t_{exp} = 0.02 s) from which the velocity components are subsequently calculated. The principle of the method can be explained by considering the time and space sequence of the travel of an isolated particle, as shown in Fig. 2(a). For the purpose of this explanation, the particle is assumed to have a constant velocity and a straight-line trajectory in the image plane. Initially, i.e. when the shutter opens, the particle is at location A. The particle travels along the path shown in Fig. 2(b) and it reaches point B by the time the shutter has closed. A laser sheet was used for convenience in the present work, but a third sheet of light of different colour could also be utilised for the same purpose.

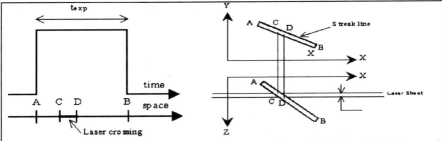

Figure 2. (a) Time and space sequence of particle travel; (b) Path travelled by a particle in the x-y and the x-z plane.

From the (x,y) coordinates of points A and B, the in-plane velocity components can be found;

$$U = \frac{x_B - x_A}{t_{exp}} \tag{1}$$

$$V = \frac{y_B - y_A}{t_{exp}} \tag{2}$$

To quantify the velocity component in the z-direction, i.e. orthogonal to the image plane, the z-coordinate of points A and B has to be determined. As the particle crosses the laser sheet at points C and D, then if the thickness of the laser sheet (Δ) can be determined, then:

$$\frac{x_D - x_C}{x_B - x_A} = \frac{y_D - y_C}{y_B - y_A} = \frac{z_D - z_C}{z_B - z_A} = \frac{\Delta}{z_B - z_A} \tag{3}$$

and the orthogonal velocity component can be determined from:

$$W = \frac{z_B - z_A}{t_{exp}} = \frac{\Delta}{t_{exp}} \frac{x_B - x_A}{x_D - x_C} = \frac{\Delta}{t_{exp}} \frac{y_B - y_A}{y_D - y_C} \tag{4}$$

The above relation shows that W can be expressed as a function of Δ, the x and y coordinates of points A, B, C and D, and t_{exp}. Whilst Δ and t_{exp} can be easily determined, the accurate determination of the x and y coordinates of points A, B, C and D necessitated the formulation of appropriate software algorithms which could account for all possible particle tracks and for any location of points A and B within the light sheets. The image processing algorithms will be described in details in another communication due to space limitations.

The inherent limitations of the technique stemming from the above methodology are two. First, it is not possible to evaluate a purely orthogonal velocity, i.e. when $U = V = 0$, as in the case of a particle crossing the laser sheet at right angles to the image plane. Second, the smallest measurable orthogonal velocity component with the *current* system is:

$$w_{min} = \frac{\Delta}{t_{exp}} = \frac{1.63 \times 10^{-3}}{20 \times 10^{-3}} \approx 0.08 \; m/s \tag{5}$$

whilst the largest measurable orthogonal velocity component depends on the thickness of the laser sheet.

Characteristic results and validation

From the 2147 images analysed, 11506 velocity vectors were determined. Calculations were performed by averaging over a 2mm × 2mm area, i.e. the measurement area of (14×18) mm^2 was divided into 63 squares, each 4mm^2. No further processing or filtering of the resulting vectors was made. The average number of velocity vectors obtained in each square was 182. It should be noted that this level of vector count per area is relatively low and its effects on measurement accuracy are highlighted below. The resulting vector map is shown in Fig. 3 (a). There is scatter in the results but the overall distribution, i.e. the peak values near the centre of the inclined jet and a drop in magnitude at y values around 0-5mm, can be observed.

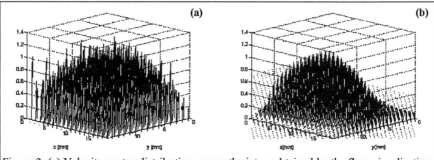

Figure 3. (a) Velocity-vector distribution across the jet as obtained by the flow visualisation technique; (b) Velocity-vector distribution across the jet as obtained by PIV.

To assess and validate the technique, measurements of the jet flow were also made with 3-D PIV system produced by Optical Flow Systems and comprising two high sensitivity 12-bit frame-straddling cameras, model PCO SensiCam. The velocity vector distribution obtained with the PIV is shown in Fig. 3 (b). The corresponding vector distribution is similar to that in Fig. 3 (a) but shows two differences: the scatter in the data is substantially lower than with the present technique due to the higher spatial resolution and higher data count of the PIV method and the values measured near the image edges, at $x = 0 - 5$mm and $y = 0 - 5$mm, are lower.

The three mean velocity components with respect to the plane of observation (U, V and W, normalised with the jet bulk velocity, Ub = jet flowrate/jet cross-sectional area = 1.15m/s), obtained with the two methods at the $y = 2$mm plane, are compared in Fig. 4 below. In Fig. 4 and Fig. 5, the dimension x is normalised by the diameter of the jet ($d = 10$mm). The V component (the near-zero value component in the image plane) profiles obtained by the two methods show excellent agreement and are similar to within $0.017Ub$ at all locations. The profiles of the second, non-zero, in-plane component, U, are also very similar but their differences are higher, up to $0.043Ub$. The out-of-plane component (W) profiles are similar to within $0.05Ub$ near the peak region (at $x/d = 0.2 - 0.8$) but show larger differences in the edges of the measurement region. The lower accuracy in the out-of-plane component (W and also w as mentioned below) measured with the 3-D visualisation technique is mainly due to the spatial variation of the laser sheet thickness (which had a maximum of 18.4% for the presented experimental configuration). The increased discrepancy between the results of the two techniques for $x/d > 1.1$ is also due to the drop in the vector count per area, as less than 182 vectors were recorded per image area for $x/d \geq 1.3$.

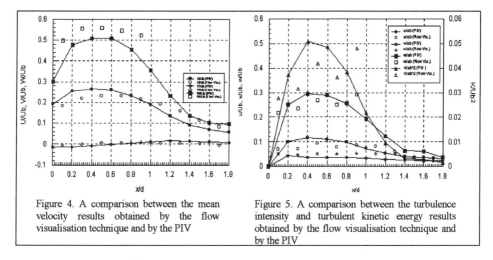

Figure 4. A comparison between the mean velocity results obtained by the flow visualisation technique and by the PIV

Figure 5. A comparison between the turbulence intensity and turbulent kinetic energy results obtained by the flow visualisation technique and by the PIV

The associated turbulence levels or RMS velocities (u, v and w) and turbulent kinetic energy (K) obtained with both the 3-D technique and PIV were calculated from the velocity values recorded and are compared in profile form in Fig. 5 for the plane $y = 0$mm. The fluctuating velocities and K have been normalised by Ub and Ub^2 respectively. The u and v profiles measured with the two techniques are very similar; differences in the v and u distributions do not exceed $0.026Ub$ and $0.034Ub$ respectively. The w profiles are similar to within $0.034Ub$ in the $x/d = 0.2 - 0.8$ region, but exhibit similar trend to that of W near the edges of the measurement region. The K profiles show more scatter due to the accumulated error in this measurement due to the individual errors in u, v and w; differences between PIV and 3-D visualisation data vary from $0.0039 - 0.024Ub^2$, with an average discrepancy of $0.012Ub^2$.

Direct comparison of the present flow data with earlier investigations of jet flows is not possible, as the present jet flow is both inclined and affected by the proximity of the confining walls and the geometry of the enclosure into which the jet discharges. The authors are not aware of measurements in geometrically similar configurations. Consequently, only tentative, approximate comparisons may be made. For example, the peak values of K and of the out-of-plane RMS component (w), are comparable to those measured with LDA by [2]: the measured K/Ub^2 and w/Ub values were around 0.08 and 0.34 respectively near the jet outlet, compared with the corresponding values of 0.05 and 0.27 at $x/d = 3.33$ and for an inclined jet with a non-zero surrounding fluid velocity in the present work.

The above results have shown that the novel 3-D flow visualisation technique can provide the in-plane mean velocity and turbulence level components with good accuracy in comparison to the PIV data. The discrepancies with the PIV data are higher for the out-of-plane component measurements and for this reason the accuracy of the 3-D technique is analysed in the following section.

Assessment and validation of the technique

Before assessing the accuracy of the new technique, it is useful to quantify the measurement errors in the 3-D PIV method against which the results are compared. Based on the extensive error analysis of [3] the error in the in-plane velocity component (measured by the present 3-D PIV system) is around 1%, whereas that in the out-of-plane component is around 4%. These errors should be kept in perspective when assessing the 3-D visualisation errors. There are two main sources of error associated with the new 3-D visualisation technique. The first is due to the resolution of the technique and stems from errors in measuring the time (t_{exp}) and space (length of streaklines and thickness of laser sheet) parameters. For a particle of high velocity, the combined error due to these three parameters was found to be 0.015 m/s or $0.013Ub$ for U whilst the error in W is 0.25 m/s or $0.21Ub$. The second source of errors is due to the statistical uncertainties due to low sample numbers employed, especially in the edge regions. For the evaluation of such errors, the formulae suggested by [4] were employed to calculate the statistical errors for all seven quantities, U, V, W, u, v, w and K. In all cases, the statistical sample size and accumulated errors accounted for the observed differences between PIV and 3-D visualisation profiles, except for W and w in the regions $x/d < 0.2$ and $x/d > 1$, where the number of data contributing to a measurement was very low (less than 50). The largest errors are, of course, found in W and w for the reasons stated earlier and in K, primarily due to the effect of the w contribution to this quantity.

Concluding remarks

The objective of the work reported in this paper was to introduce the principles and methodology of a novel quantitative 3-D flow visualisation technique. Characteristic results have been presented to demonstrate the ability of the new 3-D visualisation technique to evaluate all mean velocity components and associated turbulence quantities using a simple and inexpensive configuration employing a single colour CCD camera interfaced to a PC, and three sheets of light of different colour. The technique has been successfully applied to measure the flow of an inclined water jet confined in a rectangular enclosure. Comparison of the results with similar data obtained with a three-dimensional PIV system showed that the current technique can provide useful velocity information provided that sufficiently large sample sizes are used for the out-of-plane component measurements.

Future work shall be concerned with a) further optimisation of the image processing software to maximise the speed of data processing and increase the number of samples acquired per image area, b) incorporating light source pulsing so as to resolve the velocity direction and c) optimising the optics to minimise variations in light sheet thickness, in order to improve the accuracy of the technique and remove statistical and other uncertainties.

References

[1] Fentiman, N.J., Lee, K.C., Paul, G.R. and Yianneskis, M., The trailing vortex structure around hydrofoil impeller blades. *Transactions of the Institution of Chemical Engineers, Part A, Chemical Engineering Research and Design*, 77 (1999), 731-740.

[2] Hussein, H.J., Capp, S. and George, W.K., Velocity measurements in a high Reynolds-number, momentum-conserving, axisymmetric, turbulent jet. *J. Fluid Mech.*, 258 (1994), 31-75.

[3] Lawson, N.J. and Wu, J., Three-dimensional particle-image velocimetry: experimental error analysis of a digital angular stereoscopic system. *Meas. Sci. Technol.*, 8 (1997), 1455-1464.

[4] Yanta, W.J., Turbulence measurement with a laser velocimeter. Naval Ordnance Laboratory, Maryland, USA. Report No. 73-94 (1973).

Inst. Phys. Conf. Ser. No. 177
Paper presented at 1st Int. Conf. on Optical & Laser Diagnostics, London, 16–20 Dec. 2002
©*2003 IOP Publishing Ltd*

The Development and Application of Time Resolved PIV at The University of Strathclyde

M Stickland*, T Scanlon*, T Vidinha*, W Dempster*,
R Jaryczewski**

* Department of Mechanical Engineering, University of Strathclyde, Glasgow.
** Dantec Dynamics Ltd, Royal Portbury, Bristol.

Abstract. This paper describes the development of time resolved particle image velocimetry (PIV) within the Department of Mechanical Engineering at the University of Strathclyde. The Department's first PIV systems were developed on a limited budget and used existing and second hand equipment. The original technique which, employed 16mm high speed cinematography, is described. The introduction and development of low cost systems employing high speed digital video (HSDV) is discussed and, finally, the Department's new time resolved PIV system, supplied by Dantec Dynamics, is introduced. For each of the PIV systems that have been developed a critical analysis of their functionality is given and samples of the data that they have been produced are shown. Data are presented from systems such as de-rotated centrifugal impellers, air bubbles growing in columns of water, pulsatile jets and vortex shedding.

1. Introduction

1.1 Bubble Flows

Researchers in the Energy Systems Division of the Department Of Mechanical Engineering at the University Of Strathclyde were first drawn to the application of time resolved PIV because of their study of bubble formation, growth and detachment as part of a funded research program into the loss of coolant accident (LOCA) which could occur within a pressurised water reactor (PWR)[1], figure 1.

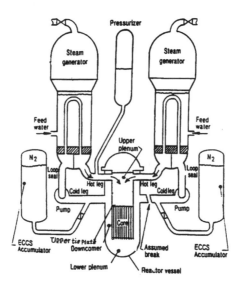

Figure 1: PWR Schematic

During a LOCA a breach within the cold leg of the PWR may occur and cause the primary coolant within the reactor to be forced out of the system. To facilitate the required heat dissipation cold water is flushed into the core from the emergency core cooling system (ECCS) accumulators. The rate at which this cooling water can enter the reactor through the hot legs is limited by the steam being generated within the core trying to force its way out through the hot legs. Steam continues to be generated until there exists a pool of water on the upper tie plate in which the steam, generated in the core, condenses. The depth of the pool and the liquid thermal conditions required to completely condense the steam generated is dependent upon the rate at which the steam is generated and the size and shape of the bubbles.

Initial research was undertaken to study and model the growth, detachment and heat transfer characteristics of steam bubbles in water. This utilised a shadowgraph technique to capture two dimensional images of the bubbles by a high speed cine camera as they developed.

It was decided to undertake some flow measurement around a growing bubble by PIV. Because the bubble growth is a highly transient phenomena the standard PIV method of taking single autocorrelation images with a CCD or 35mm camera was inappropriate as they would simply supply snapshots of the whole process. It was therefore decided to utilise high speed cine as this would capture a complete cycle of growth and detachment and allow for the flow field to be determined at a later stage It was considered that the initial application of this technique should be applied to an air bubble growing in water for the sake of safety and simplicity.

The particles in the flow were illuminated by a laser sheet produced by a 5W Spectra Physics 165 Argon Ion laser and cylindrical lens. Typically only 2.5W of laser power were required for imaging purposes. The process of bubble formation was filmed by an Hitachi 16HM high speed camera at frame rates of between five hundred and a thousand frames per second. Air flow was provided by a Charles Austin

CAPEX 2 membrane air pump at rates between 200 and 1100 cm^3 per minute measured by a rotameter. The water column was 150mm square section and 300mm high manufactured in glass with a 4mm orifice at the base. A diagram of the experimental setup is shown in figure 2.

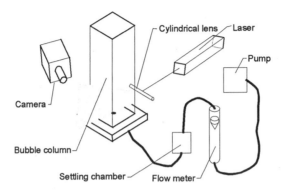

Figure 2: Experimental setup

Images were digitised by projecting the 16mm negative image onto a screen and grabbing with a Sony XC75 768 x 494 pixel monochrome CCD camera connected to a Matrox frame grabber card in a 486SX25 PC. The digitised images were analysed by Optical Flow Systems VidPiv PIV analysis Software by cross correlation on a 486DX266 PC. A typical image may be seen in figure 3.

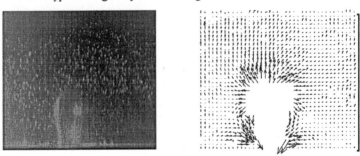

Figure 3: Original bubble PIV data

The process was extremely time consuming and took several days between taking the images and producing the vector map. To speed up the process a high speed digital video was purchased. The Kodak Motioncorder was one of the first relatively low cost, approximately £10k, high speed digital videos. However, whilst it stored the data digitally it could only provide an analogue, PAL, output.

The process of bubble formation was now filmed by the Kodak Motioncorder which had a capability of recording at frame rates up to 600 frames per second. The images were digitised by a Matrox frame grabber card in a 486DX266PC. The digitised images were analysed by Optical Flow Systems VidPiv PIV analysis Software by cross correlation on a Pentium P200 PC. Typically, the images were acquired at 240 frames per second and a resolution of 256x256 pixels. The time between taking the images and producing a single vector map had reduced to approximately an hour [2]. The data was of good quality and was used to validate some simple, two dimensional, inviscid models of the flow field, figure 4.

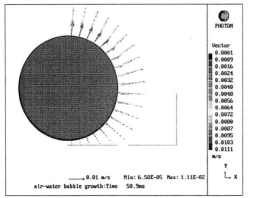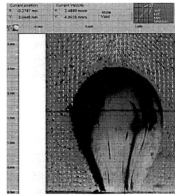

Figure 4: PIV image and vector map acquired by Kodak Motioncorder

In early 2000 a Photron Fastcam super 10KC was acquired. The Fastcam is a high speed digital video capable of capturing up to 10k images per second It has a higher resolution that the original Motioncorder and can digitally down load directly to a PC via a SCSI 2 link in .bmp format. Illumination was still provided by a 5W Spectra Physics Argon Ion laser and analysis was by cross correlation using VidPIV V2.41 from Optical Flow Systems of Edinburgh. The time from acquisition to production of a single vector map had now reduced to just ten minutes.

2. Time Resolved PIV system

The quality of the images and data acquired by the original systems were remarkable considering the rather haphazard way that the system had been developed over a number of years. In 2000, a grant from the Engineering and Physical Sciences Research Council allowed the purchase of a specifically designed system.

The system acquired from Dantec Dynamics consisted of a diode pumped Nd:YAG Laser, high speed digital video (HSDV), Flowmanager analysis Software and TimeResolve trigger and synchronistion system. The Nd:YAG laser is capable of pulse repetition rates up to 50 kHz with 5mJ per pulse at 10 kHz. With a pulse width of approximately 190 ns. It requires a 25A single phase power supply and has an integral chiller unit on the power supply which does not require an external water supply for cooling. The HSDV has a maximum resolution of 1024x1024 pixels at frame rates up to 500 frames per second reducing to 512x128 at 4000 frames per second. The TimeResolve synchronisation system allows the laser and camera to be slaved together to run in either single shot mode or double pulse, frame straddling, PIV mode. Figure 5 shows the two modes in which the laser may be triggered.

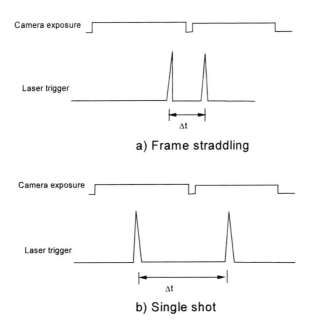

a) Frame straddling

b) Single shot

Figure 5: Camera and Laser trigger sequencing

2.1 Single shot mode.

In the single shot mode the laser is synchronised to fire a single 190 ns pulse in the centre of each camera exposure. The time delay between images is therefore regulated by the exposure rate of the HSDV – from 16 ms to 25μs. This allows the cross correlation of images for relatively low speed flows at high resolution but for higher speed flows the resolution will, necessarily, be reduced due to the higher frame rate of the camera. For high resolution at higher speeds frame straddling was required.

2.2 Frame straddling – PIV mode

In the frame straddling mode the laser is triggered just before the end of the first exposure and shortly after the start of the second exposure of the HSDV. In this mode it is possible to vary the time between exposures from 20μs seconds up to 200μs seconds allowing the measurement of relatively fast flows. As the time delay is now independent of the camera frame rate high resolution images may be acquired.

2.3 Laser power regulation

In the frame straddling mode it is necessary to control the power in each laser pulse. As the pulses are not regularly spaced the first pulse will contain an excessive amount of power whilst the second pulse, having less time for the laser to charge, will be relatively weak. Also, in single shot mode, care must be taken not to run the laser at full power at low repetition rates (less than 5kHz) as this may damage the second harmonic generator (SHG) crystal. The TimeResolve software from Dantec ensures that SHG crystal will not be damaged at any pulse rate.

3.0 Sample data

Below are some images and the associated data that have been collected by the system. The data are just small selections from much larger data sets. Animations of theses data sets may be found on the accompanying CD ROM or by visiting htttp:\\www.homepages.strath.ac.uk\~clcs20\

3.1 Bubble Formation.

Figure 6 shows the vector field surrounding an air bubble growing from a submerged orifice. The orifice diameter was 6mm and the air volume flow rate 90 cc/min. The data was acquired at 1000 frames per second in single shot mode. Typically, between 50 and 100 vector maps were acquired for each bubble formation [3].

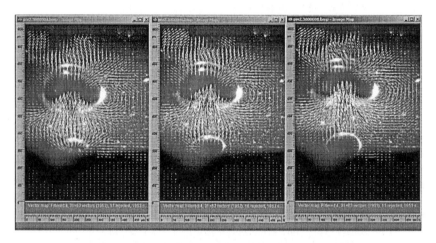

Figure 6: Vector field around a bubble growing at a submerged orifice

3.2 Derotated impeller flows

The flow patterns between the blades of an impeller can be turbulent and constantly changing with time. This turbulence will lead to poor performance of the impeller and, if excessive, may lead to the early failure of the impeller or to the creation of unacceptable noise. It is, therefore, to the impeller designers advantage to discover the nature and cause of these secondary flows so that they might be eliminated.

Analysis of secondary flows is difficult when the observer remains stationary relative to the impeller as they are hidden within the dominant primary flow. If, however the observer is able to rotate in the plane of the impeller at the same rotational speed then the impeller appears to be stationary, the primary flow disappears and the observer only sees the secondary flow field relative to the stationary impeller.

The authors have been able to obtain the relative view of a flow field in an impeller by the use of an image de-rotator. The image derotator optically rendered completely stationary the image of the rotating impeller. The derotated view of the impeller was then analysed by particle image velocimetry to provide the relative velocity field within the radial blade passages of the centrifugal impeller [4][5].

Water was used as the working fluid, seeded with 50µm diameter polyamid particles. Data was acquired at 1000 Hz in frame straddling PIV mode with a pulse separation of 100 µs. Figure 7 shows the velocity vectors on the pressure and suction sides of an impeller blade rotating at 650 rpm.

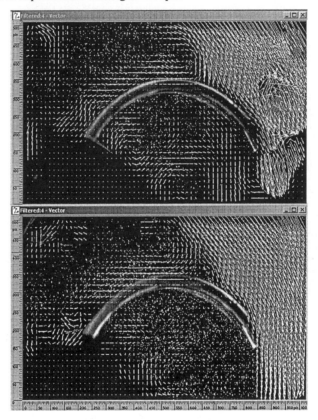

Figure 7: Velocity vectors on the pressure and suction sides of an impeller blade.

3.3 Pulsatile jet flow

Figure 8 shows the velocity vectors of a jet of fluid issuing into a tank of water. The diameter of the jet was 5 cm and the jet was pulsed sinusoidally at approximately 1Hz. The data was acquired in frame straddling PIV mode with a pulse separation of

200 µs at a frame rate of 1000Hz. The images shown were acquired just as the jet started issuing into the tank and clearly show the starting vortex being developed.

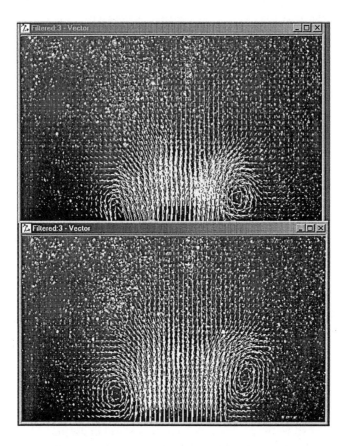

Figure 8: Pulsatile jet flow.

4. Discussion

The Department of Mechanical engineering has, for several years, persevered with the development of time resolved PIV. The original systems were rather crude and required an excessively long time to produce a limited number of data sets. However, the development of solid state pulsed lasers and digital high speed video and high speed digital computers has allowed much of the data acquisition process to become automated. The increased power of the light sources and the increase in resolution of the cameras has also allowed a significant improvement in the quality of the raw data produced. Combined with the massive increase in the computer power available in a relatively cheap PC this has allowed the analysis of the data sets to be accomplished within seconds. The development of digital cameras capable of acquiring data at over

10kHz has now even made it possible to contemplate the statistical analysis of these data sets to provide turbulence data.

The data shown in this paper are, by necessity, a small part of much larger data sets. The benefits of time resolved PIV can only truly be appreciated by studying these large data sets when the structure of transient flows can be fully appreciated.

5. Conclusions

The time resolved PIV system used by authors has successfully measured several different transient flow fields. The range of different transient flow phenomena which will benefit from measurement by this new technique is almost limitless and the possibility of utilising it for more fully understanding the nature and structure of turbulence is intriguing

References

[1.]   Stickland M T, Dempster W, Lothian L, "PIV Studies of Bubble Growth and Detachment by High Speed Photography". 22nd International Symposium on High Speed Photography and Photonics, Santa Fe, USA, SPIE, October 1996.

[2.]   Dempster Wm, Stickland M. "Particle Image Velocimetry Studies of Bubble Growth and Detachment by High Speed Video Photography".2nd International Symposium on Two Phase Flow Modelling and Experimentation, PISA, May 23-26 1999 Vol III pp 1365-1370ISBN:88-467-0177-3

[3.]   T Vidinha, W Dempster, M Stickland, "Liquid Velocity Field Measurements during Bubble Formation and Detachment at an Orifice". EALA Conference, Sept 2001, Limerick.

[4.]   M Stickland, Blanco E, J Fernandez, Parrondo J, T Scanlon, "A Numerical And Experimental Analysis Of Flow In A Centrifugal Pump" The 2002 Joint US ASME-European Fluids Engineering Summer Conference. July 14-18, 2002, Montreal

[5.]   M Stickland, J Fernandez, T Scanlon, P Waddell, "Mapping The Velocity Field In A Pump Impeller Using High speed digital video, Image de rotation and Particle Image Velocimetry". EALA Conference, Sept 2001, Limerick

Inst. Phys. Conf. Ser. No. 177
Paper presented at 1st Int. Conf. on Optical & Laser Diagnostics, London, 16–20 Dec. 2002

Velocity Field Measurements in a Non-Premixed, Bluff-Body Burner using Digital Particle Image Velocimetry

G K Hargrave, R Carroni* and H K Versteeg

Wolfson School of Mechanical and Manufacturing Engineering, Loughborough University, Loughborough, LE11 3TU, UK.
*Presently at: ALSTOM Power Switzerland CH-5405 Baden-Dättwil, Switzerland.

Abstract. The non-premixed, bluff-body burner provides both an industrially significant flow configuration and an academically useful tool for the development and validation of predictive CFD models. The bluff-body configuration is used to provide flame stabilisation in burners ranging from a few kW to several MW. The unsteady nature of the flame/flow interaction in the recirculating flow of a bluff-body wake provides a challenging scenario for CFD. Previous studies of bluff-body stabilised flows have provided detailed measurements of velocity fields for unconfined flames only. The aim of the current work is to provide detailed data of the velocity field in a laboratory scale, confined, non-premixed bluff-body burner with a blockage ratio of 0.44. The fuel and air streams were introduced into a circular combustion chamber of 84.5mm internal diameter manufactured from transparent Suprasil to allow the application of laser-based optical diagnostics. Data is presented for the velocity field for reacting flow measured using digital particle image velocimetry (DPIV) consisting of 2-D maps of time-averaged velocity and turbulence fields together with derived profiles of turbulence energy and shear stress. The prevailing fuel and air flow rates gave rise to a flow pattern that was dominated by three toroidal recirculation zones. The influence of these recirculation zones on the flow behaviour is investigated.

1. Introduction

The non-premixed, bluff-body burner provides both an industrially significant flow configuration and an academically useful tool for the development and validation of predictive CFD models. In practical combustion devices, the bluff-body configuration is used to provide flame stabilisation in a wide range of burner sizes from commercial scale combustors of just a few kilowatts to industrial scale burners of several megawatts. In terms of CFD modelling, the unsteady nature of the flame/flow interaction in the recirculating flow of a bluff-body wake provides a challenging scenario for physical sub-model development. A review of public domain literature revealed that there is a paucity of experimental velocity measurements for confined bluff-body stabilised flames, particularly those at high blockage-ratio where flow effects due to confinement are most likely to be significant. Previous studies of bluff-body stabilised flows have provided detailed measurements of temperatures and mixture fractions by Correa and co-workers [1,2] for a confined axisymmetric bluff-body combustor with a low value of the blockage ratio of 0.125. Roquemore *et al* [3] provided

LDA flow field data for a similar burner where the confinement effect is limited. Detailed data including velocity fields are only available for unconfined flames [4-8]. Investigations into the effect of confinement by Schefer *et al* [9] included no velocity field data.

In this paper we present velocity fields, measured using digital particle image velocimetry (DPIV), and radial temperature profiles in a laboratory scale, confined, non-premixed bluff-body burner with a blockage ratio of 0.44 tested under low Reynolds number conditions. We discuss the main features of the complex flow field with references to experimental 2-D maps of time-averaged velocity and turbulence fields as well as derived radial profiles of turbulence energy and shear stress.

2. Experimental Procedure

Burner. A schematic of the burner is shown in Figure 1. The burner was designed to achieve fully developed flow conditions at the air and fuel inlets. It consisted of a 4mm diameter fuel jet ejecting into the wake of a circular bluff-body of 44.5mm diameter mounted concentrically in an air channel with a diameter of 67mm, giving a blockage ratio of 0.44.

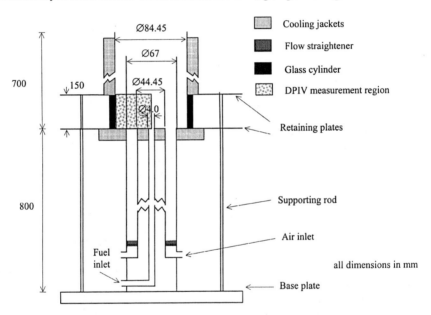

Figure 1: Schematic of bluff-body burner with main dimensions

The fuel/air flow issued into a cylindrical combustion chamber of 84.5mm internal diameter and length of 700mm, the first 150mm of which was manufactured from transparent Suprasil to allow optical access. Technical grade methane (100% purity) was used as fuel and annulus air was taken from the laboratory compressed air mains. Fuel and air flowrates (4 and 480 litres/min at s.t.p. conditions, respectively) were set to generate a ratio of mean fuel to air velocity around 1.3. This is within the range associated with flow patterns that are dominated by large toroidal vortices attached to the bluff body, which are understood to be involved in flame stabilisation. The fuel and air inlet ducting was configured to provide a substantial upstream straight length (100 diameters and 35.6 annular gap widths, respectively), which ensures that the inlet flow into the combustion chamber is close to fully developed. Temperature measurements were made with 0.8mm diameter K-type thermocouples, which were mounted on a radial traverse to acquire profiles at axial locations x/R_b = 0.5 – 4.5 in increments of 0.5, where R_b is the outer radius of the bluff-body.

Optical diagnostics. Digital Particle Image Velocimetry (DPIV) was used to generate two-dimensional quantified flow maps. A diagram of the DPIV system is given in Figure 2.

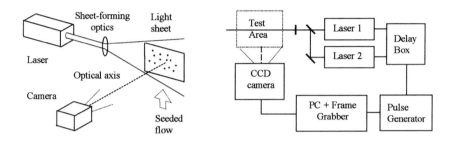

Figure 2: DPIV system

The air supply flow was seeded with zirconium oxide particles with a nominal diameter of 1μm. Two lasers (Continuum Surelite II, 532nm Nd:Yag, 200mJ) were used in conjunction with a delay mechanism to provide illumination of the seeded flow with a time delay of 100μs. Sheet optics were introduced to generate two laser light sheets (approximately 100mm wide and less than 1mm thick), which were positioned such that the first region to be investigated was the recirculation zone immediately behind the bluff-body face. Nearby regions could be studied by moving the burner, which was placed on a specially adapted traversing mechanism permitting horizontal and vertical motion nominally in the plane of the laser light sheet. A CCD camera (Kodak Megaplus ES1.0, 1000×1000 pixels) recorded a square flow region of 40mm×40mm giving a spatial image resolution of 40.8μm/pixel. The camera was linked to a frame grabber, capable of recording sequences of 180 particle image pairs. The instrumentation was triggered by a pulse generator. The image pairs were used to generate mean and fluctuating velocity fields using cross-correlation analysis with the software package Visiflow using 64 by 64 pixel interrogation regions. These overlapped by 50%, which provided 2.6mm square interrogation regions spaced at 1.3mm. The mean quantities U and V and fluctuating components $\overline{u'^2}$, $\overline{v'^2}$ and $\overline{u'v'}$ were computed along selected radial traverses. This was accomplished by averaging information from 180 consecutive images, resulting in radial profiles of the above quantities at axial locations x/R_b = 0.5 - 5.0 in increments of 0.5 (R_b = bluff-body outer diameter).

Uncertainty estimates. The positional uncertainty associated with the traversing mechanism was estimated to be ±0.5mm and the uncertainty of flow rate setting was ±0.4% for fuel and ±0.8% for air. The worst-case estimate of the uncertainty associated with the DPIV measurement of mean velocities ±4% was based on an assumption of particle displacement that is accurate to within ±0.5pixel. It should be noted that the central fuel jet was unseeded in these experiments, which causes velocity bias problems near the centreline close to the fuel inlet. Mean and fluctuating velocity values should be treated with caution in this region. The maximum uncertainty of the temperature measurement was estimated to be around ±20% due to systematic bias associated with radiative heat exchange and soot effects. Further details of the experimental design and procedure can be found in Carroni [10].

3. Results and Discussion

Figure 3 shows the visible envelope of the flame, which is seen to be lifted off the bluff-body face. It has an approximate height of 3.3 bluff-body diameters and is slightly wider than the bluff-body. Figure 4 shows a composite vector plot based on instantaneous data from a number of adjacent regions and clearly reveals the complex flow structure of the burner. The principal feature is the large recirculation region, marked A, with an axial length around $2.8R_b$. This is driven by the air jet B, which emerges from the burner annulus. The central fuel jet is found to persist up to $x/R_b \approx 3.5$. Figure 4 also suggests the presence of a number of smaller eddy structures between recirculation zone A and the centreline. The most distinct structure is marked C and occurs near the bluff-body extending in the axial direction up to $x/R_b = 1$. Finally, a clearly-defined recirculation area, marked D, with an axial extent of $x/R_b = 1.3$ occurs behind the step between the air inlet and the outer wall of the burner. It is interesting to note that in unconfined burners (see e.g. Ref.[8]) the annular air flow is deflected towards the centreline. In our burner, however, the air jet is initially deflected towards the outer wall of the burner by the interactions between recirculation zones A and D. The air stream only starts to move inwards at the top of recirculation zone A at $x/R_b \approx 3 - 3.5$.

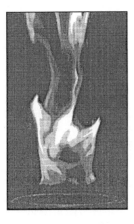

Figure 3: Visible flame structure

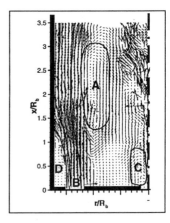

Figure 4: Sample instantaneous velocity field

Figures 5a-f shows radial profiles of $U, V, \overline{u'^2}, \overline{v'^2}, \overline{u'v'}$ and T. At $x/R_b = 0.5$ the mean axial velocity component U is highest just above the air annulus, as expected. Large values of the mean radial velocity component V around $r/R_b = 1-1.5$ at $x/R_b = 1.5$ and 2.5 are associated with recirculation zone A. Due to the velocity bias problems noted earlier the centreline velocity values in the vicinity of the burner face are suspect. However, at $x/R_b = 4.5$ mixing between the fuel and air streams yields adequate seeding levels and the results indicate that the fuel jet has been broken down. Figures 5c-d show high levels of the fluctuating velocity components in the near-burner region $r/R_b < 0.5$ and $x/R_b = 2.5 - 3.5$, which are of the same order of magnitude mean velocity U. These are probably associated with interactions between recirculation region A and the fuel jet. Figure 5e suggests that regions of high turbulent shear stress $\overline{u'v'}$ broadly appear in the vicinity of regions with high mean velocity gradients.

Our temperature data in Figure 5f shows that the peak temperatures occur at $r/R_b \approx 0.9$ and $x/R_b = 1.5 - 3.5$. Thus, Figures 4 and 5f show that the hottest region is located inside the main recirculation zone A, indicating that the bulk of the air/fuel mixing takes place here. Moving outward in the radial direction, the temperature decreases rapidly due to cooling by incoming annulus air. In the centreline region the temperature profile becomes progressively flatter as x/R_b increases. We note that $x/R_b = 4.5$ is located just above recirculation region A. Figure 4 suggests that strong bulk convection causes movement of hot reaction products and colder annulus air towards the centreline, where further mixing produces the flattening of the temperature distribution.

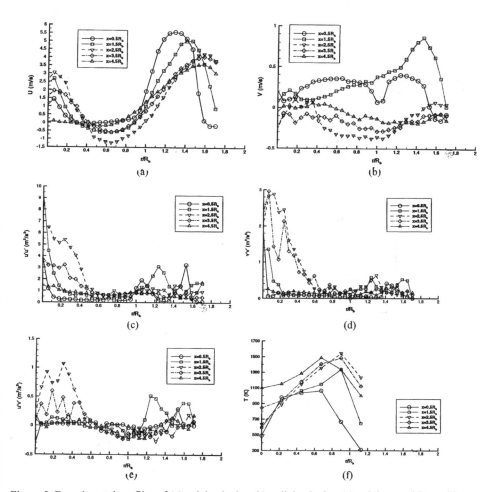

Figure 5: Experimental profiles of (a) axial velocity, (b) radial velocity, (c) axial normal Reynolds stress, (d) radial normal Reynolds stress, (e) Reynolds shear stress, (f) temperature

4. Conclusions

We have presented velocity field date measured with DPIV complemented by a selection of radial temperature profiles inside a laboratory scale, confined, non-premixed bluff-body burner with a blockage ratio of 0.44. The flow conditions ($V_{fuel}/V_{air} \approx 1.3$) were selected to generate conditions corresponding to flow patterns that are dominated by large toroidal vortex regions behind the bluff-body leading to flame stabilisation. Experimental 2-D flow maps revealed the existence of a complex flow field containing three recirculation regions along with an annular air jet flow. In our burner, the air jet is initially deflected towards the burner walls by the interplay between two of the recirculation zones. Only much further downstream the air stream moves inwards towards the centreline, as in unconfined burners. Radial temperature distributions showed that the combustion reaction mainly takes place in the largest of the recirculation regions where fuel and air are brought together. Detailed radial distributions of mean and fluctuating velocity were given to facilitate comparison with CFD results for this challenging combusting flow scenario.

References

[1]   Correa SM and Gulati A 1992 *Combustion & Flame*, **89**, 195-213.
[2]   Correa SM and Pope SB 1992 *Proc. 24th Symp. (Int.) on Combustion*, The Combustion Institute, 279-285.
[3]   Roquemore WM, Bradley RP, Stutrud JS, Reeves CM and Krishnamurthy L 1980, *Proc. ASME Gas Turbine Conf. & Prod. Show*, New Orleans, LA.
[4]   Schefer RW, Namazian M and Kelly J 1987 *Combustion Sci. & Tech.*, **56**, 101-138.
[5]   Schefer RW, Namazian M and Kelly J 1994 *AIAA J.*, **32**, 9, 1844-1851.
[6]   Masri AR, Dibble RW and Barlow RS 1992 *Proc. 24th Symp. (Int.) on Combustion*, The Combustion Institute, 317-324.
[7]   Masri AR, Dibble RW and Barlow RS 1996 Prog. Energy Combustion Sci., **22**, 307-362.
[8]   Masri AR, Kelman JB and Dally BB 1998 *Proc. 27th Symp. (Int.) on Combustion* The Combustion Institute, 1031-1038.
[9]   Schefer RW, Namazian M, Kelly J and Perrin M 1996 *Combustion Sci. & Tech.*, **120**, 185-211.
[10]   Carroni R 1999 *PhD Thesis*, Loughborough University, Loughborough, UK.

Inst. Phys. Conf. Ser. No. 177
Paper presented at 1st Int. Conf. on Optical & Laser Diagnostics, London, 16–20 Dec. 2002
©2003 IOP Publishing Ltd

2-D LIF measurements of thermo-acoustic phenomena in lean premixed flames of a gas turbine combustor

Sabine Schenker† , Rolf Bombach†, Andreas Inauen†, Wolfgang Kreutner†, Walter Hubschmidt†, Ken Haffner‡, Christian Motz‡ Bruno Schuermans‡, Martin Zajadatz‡ and C Oliver Paschereit‡

† Combustion Research, Paul Scherrer Institut, CH-5232 Villigen PSI, Switzerland

‡ ALSTOM (Schweiz) AG, CH-5405 Baden-Dättwil, Switzerland

Abstract. Thermo-acoustic phenomena manifest in lean premixed flames of gas turbines as pressure oscillations at distinct frequencies characteristic of the burner design and its operation. They can lead to early materials aging or even severe damages. Therefore, a thorough understanding of the processes responsible for these instabilities, i.e. the coupling between the unsteady heat release and the pressure fluctuations, is crucial. In order to study these instability modes, phase-locked 2-D OH laser-induced fluorescence (LIF) measurements have been performed. The fluorescence experiments were carried out on a test rig equipped with a commercial 700 kW burner and a combustion chamber of UV transparent quartz, using a pulsed Nd:YAG/dye laser system and an intensified CCD camera for detection. Intensity variations in the integral OH LIF signal of up to ± 15% over one oscillation period are observed for peak sound pressure of 6 mbar and more. In addition, the phase-averaged position of the flame zone varies in axial direction, i.e. the main flow direction. The outer part of the flame zone, close to the combustor walls, shows much stronger oscillations than the central part. The findings indicate two counterpropagating recirculation zones - one in the center and one close to the tube walls - in agreement with CFD calculations and water channel experiments.

1. Introduction

Thermo-acoustic phenomena are of major concern for gas turbine manufacturers. The pressure fluctuations involved lead to increased pollutant, i.e. NO_x formation due to inhomogeneous temperature distributions. Furthermore, they can cause severe mechanical stress decreasing the lifetime of the combustors considerably. These combustion instabilities can be controlled either passively or actively and several techniques have been developped for the latter. Most of them base on acoustic excitation inside the combustion chamber or on fast actuators regulating either the main fuel injection or the pulsed injection of additional secondary fuel.

Whitelaw and collaborators [1] report active instability control (AIC) experiments on a small-scale methane combustor with a maximum heat release of 150 kW. Some experiments were carried out with pulsed injection of methane, i.e. the main fuel. However, the most successful damping of the combustion instabilities was achieved by oscillating injection of secondary fuel. They used two different actuators allowing either regulation of the fuel injection or pulsed ignition of the fuel/air mixture. Both AIC methods reduced the oscillations of 7 to 10 kPa by 12 to 16 dB.

Hermann et al. [2] developped an AIC method for liquid fuel spray combustors. They derived a transfer function between the observed pressure oscillations and the modulation of the liquid fuel flow rate using a piezo actuator. In addition, they made use of the acoustic resonance in the fuel supply leading to a reduction of the sound pressure amplitude by up to 40 dB.

Mokhov et al. [3] performed phase-locked CARS temperature measurements in order to determine the interaction between the acoustic oscillations and the temperature in laminar premixed flames. Based on the correlations found, they modulated the heat exchange between dame and burner surface, thus, inducing periodic changes in the flame temperature. Using a propane dame of fuel/air equivalence ratio ϕ of 0.9 they were able to vary the flame temperature by up to 300 K.

Neumeier et al. [4] examined the secondary combustion process response (SCPR) produced by a gaseous fuel injector actuator. The SCPR process was locked both to pressure data and to direct measurements of the radiation of the global combustion zone. Both triggering methods agree well and show a potential for instability control in the frequency range between 0 and 800 Hz.

Paschereit and collaborators [5-9] studied the structure and control of thermoacoustic instabilities in gas turbine combustors. They looked at various burner types for lean premixed combustion in the scale up to commercial ones of 700 kW. Combustion instabilities were either induced by appropriate operation conditions or forced by external loudspeakers.

We performed phase-locked 2-D laser-induced fluorescence (LIF) of the hydroxyl radical (OH) and of acetone in order to investigate the coupling between the heat release and the acoustics in the combustor and its feedback to the fuel/air premixing in the plenum chamber, respectively. For comparison purposes, the chemiluminescence of the flame was recorded as well.

Part of the investigations described in this article were performed in collaboration with an industrial partner, resulting in limitations on reporting experimental and technical details of the burner operation and construction.

2. Experimental

All measurements were carried out in a commercial premixed 700 kW gas turbine combustor of ALSTOM, driven by natural gas. Parts only of the different sections of the test

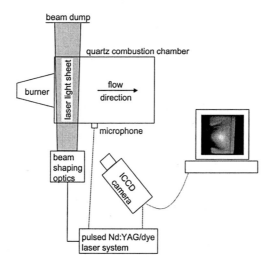

Figure 1. Schematic arrangement of the experimental setup.

rig, i.e. the burner itself (left side) and the quartz combustion chamber are schematically drawn in Figure 1. The fuel and air supply upstream of the burner as well as the stainless steel exhaust are not shown. For most experiments an orifice plate was mounted at the end of the exhaust in order to reduce the reflection of the sound back into the burner chamber [8]. To measure the sound pressure in the combustion chamber, a microphone was positioned close to the flame zone. Often, the sound generated by the flame turned out to be of too transient nature for phase-locked measurements. Therefore, two sets of four loudspeakers each were mounted around the fuel/air supply upstream of the burner and at the end of the exhaust to force the phase locking. Although the power of the eight loudspeakers (8 x 150 W) was low compared to the sound generated by the flame, a sufficiently strong forcing yielding a steady sound level could be achieved with this arrangement.

The frequency doubled output of a pulsed Nd:YAG/dye laser system at approximately 286 nm was used for excitation of the OH radicals. Due to design restrictions of the laser, its repetition rate was fixed to 20 Hz, whereas the dominant frequency of the combustion instabilities was above 85 Hz. A specially designed phase-lock loop (PLL) filter allowed for locking of the repetition rate to the specific thermo-acoustic frequency, thus yielding an interleave factor of ≥ 5. By appropriate combination of cylindrical and spherical lenses the laser beam was shaped into a thin divergent light sheet, passing from below through the combustion chamber tube, see Figure 1). The sheet width was 11 and 14 cm at the entrance into the tube and at the exit, respectively. The sheet plane was arranged parallel to the symmetry axis of the tube, i.e. the main flow direction. Using an intensified CCD camera equipped with a band pass filter centered at 308nm, the LIF signal was detected perpendicular to the plane of the light sheet. The experimental data were corrected for the above mentioned beam divergence. For acetone tracer LIF measurements of the fuel/air premixing in the plenum chamber the same optical setup has been used, except for the fourth harmonic (266nm) of the Nd:YAG laser for excitation.

3. Results and discussion

Measurements were performed for a large number of burner configurations and operating conditions. Figure 2 shows so-called phase-averaged (averaged over 100 single-pulse pictures at a fixed phase angle) images of the OH LIF intensity in the flame region of the combustion chamber for a specific burner operation. The pictures were recorded with a phase resolution of 15° over one period of the oscillation mode. Two characteristics can immediately be seen in these phase-averaged data. First, there is a temporal variation of the integral OH LIF intensity during one oscillation period. This corresponds to an oscillation in heat release, as the OH LIF intensity can be considered as a measure for the heat release. Second, there is a periodic spatial motion of the OH zone, i.e. the heat release zone.

| 0° | 45° | 90° | 135° | 180° | 225° | 270° | 315° |

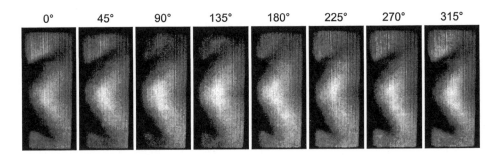

Figure 2. Phase-averaged OH LIF intensity measurements at the phases of 0°, 45°, 90°, 135°, 180°, 225°, 270ᶜ and 315ᶜ for one oscillation period.

For the further discussion, the local LIF intensity is assumed to be proportional to the local heat release. The amplitude of the sound pressure emitted by the flame can be determined from the phase-averaged integral LIF intensity as follows. For that purpose, the variation of the heat release can be put into the acoustic wave equation as driving force [10]. Neglecting any other source term, including the turbulence, and inserting the parameters of the burner operation, a sound pressure with a peak value of 3.2 mbar results for a sinusoidal LIF intensity variation of about ±15%, as observed in our measurements. This value is of the same order of magnitude as the value of 6 mbar measured by the microphone without operation of the loudspeakers. It has to be taken into account that the feeding of the sound, generated by the flame, into the standing acoustic wave in the combustion tube depends on the relative phase of the two waves. In addition, the sound propagating downstream is reflected at the tube end. Without orifice plate, the reflection of the dominant frequency amounts to about 95% (see e.g.

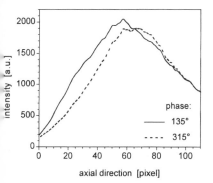
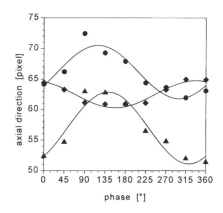

Figure 3. a) Profiles of the OH LIF intensity for the flame center along the axial burner direction for the phases of 135° (——) and 315° (- - -), respectively. b) Position of the center of mass for the OH LIF intensity in the flame zone (◆), and in the bottom (●) and top (▲) recirculation zones over one oscillation period.

Ref. [11] for a recipe of calculation), whereas with orifice, the reflection is reduced to a value of about 50% [8].

Figure 3a) shows the profiles of the OH LIF intensity for the flame center along the axial burner direction for the phases of 135° and 315°. As a measure for the spatial flame oscillation in axial direction, i.e the main flow direction, the position of the center of mass for the OH LIF intensity is defined. This quantity has been determined for averages of the central and the outer regions of the flame cross section as depicted in Figure 3b). In general, the recirculation zones of the flame front oscillate stronger and oppositely to the central part.

4. Summary

Turbulent combustion in general produces heat release that varies in time and position. The generated sound shows a broad frequency spectrum. For the combustor investigated by us, the reflection of the sound at the tube exit gives rise to a standing wave inside the tube, and it feeds the acoustic disturbance back to the region of the flame and the fuel injection. Constructive superposition of incident and reflected wave is possible only for the fundamental frequency and its overtone. The sound wave modulates the fuel concentration in the fuel injection in the plenum chamber and influences the flame speed as well. Both processes lead to variations in heat release of the flame which in turn induce sound oscillations. In this way, a high amplification of sound results.

References

[1]   Sivasegaram S, Tsai R and Whitelaw J H 1995 *Comb. Sci. Technol.* **105** 67–83
[2]   Hermann J, Gleis S and Vortmeyer D 1996 *Comb. Sci. Technol.* **118** 1–25
[3]   Mokhov A V, Aptroot R and Levinsky H B 1996 *Comb. Sci. Technol.* **120** 1–9
[4]   Neumeier Y, Nabi A, Arbel A, Vertzberger M and Zinn B T 1997 *J. Prop. Power* **138** 213–232
[5]   Polifke W, Paschereit C O and Sattelmayer T 1997 *VDI-Berichte* 455–460
[6]   Paschereit C O, Gutmark E and Weisenstein W 1998 *Comb. Sci. Technol.* **138** 213
[7]   Paschereit C O, Gutmark E and Weisenstein W 1998 *4th AIAA/CEAS Aeroacoustics Conf.* Paper AIAA 98-2272
[8]   Paschereit C O, Gutmark E and Weisenstein W 2000 *AIAA Journal* **38** 1025
[9]   Paschereit C O, Flohr P and Schuermans B 2001 *39th AIAA Aerospace-Sciences Meeting* Paper AIAA 2001-0484
[10]   Ingard U 1968 *Handbook of Physics* ed E U Condon and H Odishaw (New York: McGraw-Hill)
[11]   Kinsler L E, Frey A R, Coppens A B and Sanders J V 2000 *Fundamentals of acoustics* (New York: John Wiley)

Acknowledgments

This work was financially supported by the Swiss Federal Office of Energy (BFE).

Inst. Phys. Conf. Ser. No. 177
Paper presented at 1st Int. Conf. on Optical & Laser Diagnostics, London, 16–20 Dec. 2002
©2003 IOP Publishing Ltd

High Speed Imaging of Shocked and Detonated Gases

David A Holder, Alan V Smith, and Chris J Barton

AWE, Aldermaston, Reading, Berkshire, RG7 4PR, UK

Abstract. The successful use of a pulsed laser and 35mm streak drum camera in the recording of gas shock phenomena in shock and detonation tubes is demonstrated and discussed. One application will show a sequence of laser sheet images of the shock-induced mixing of gases of different density (one seeded). A second will show shadowgraph image capture of the fusing of detonation waves from multi-point ignition of an oxy-acetylene gas mixture. Also a novel technique combining shadowgraphy and laser sheet imaging will be presented and discussed. This allows visualisation in experiments of both detonation waves and subsequent shock-induced turbulent mixing. The limitations and relative merits of photographic film and CCD camera imaging will be briefly discussed. A video sequence shows the capability of recording 50 images on a single shock tube experiment.

1. Introduction

For a number of years shock tubes have been used at AWE to study shock induced Richtmyer-Meshkov turbulent mixing. This is to investigate the mixing across the boundaries between gases of different densities induced by the passage of a shock. These studies have been variously reported at the International Workshops on the Physics of Compressible Turbulent Mixing (IWPCTM) [1, 2, 3, 4]. The experiments are used to validate turbulent mixing codes being developed at AWE. A conventional rectangular compressed air driven shock tube, cross section 200 x 100mm, has been used for these studies. The experiments are mainly investigating the effect on the mixing process of introducing a perturbation of one of the gas boundaries, these experiments are considered to be 2D on average.

Latterly there has been a requirement to extend these studies into 2D cylindrically convergent geometry. This is being achieved through the development of a novel Convergent Shock Tube (CST) driven by detonation of an oxy-acetylene mixture. Recent developments on the CST have advanced to demonstration of a working facility capable of enabling turbulent mixing studies to be performed. Additionally it provides a means of extending experimental studies to a shock Mach Number exceeding 3, unattainable with existing linear shock tubes. The design critically demands the simultaneous ($\ll 1 \mu s$ variation) detonation of the oxy-acetylene mixture using a multi-point (30) ignition system. Earlier problems experienced with inconsistent performance stemmed largely from failure to achieve the required simultaneity conditions and the tendency for deflagration to occur rather than detonation. Such problems have been largely solved using a small optical detonation test cell incorporating the essential features of the CST oxy-acetylene gas chamber.

Pulsed copper vapour lasers and streak drum cameras are used as the main diagnostic in these experiments.

2. Laser sheet imaging of gas mixing experiment

The laser sheet technique is used routinely in the linear shock tube experiments as illustrated in figure 1. A shock with a constant pressure and flow velocity of 7ms duration is used in these experiments. The imaging system must record the incident shock passage through the dense gas region, reflection of the shock from the solid end wall of the shock tube and its return through the shocked gas region. Thus the system is set to record for 4ms. A laser sheet of 1-2mm thickness and the full 200mm height of the test cell enters through a transparent end plate. It illuminates the dense gas region which has been seeded with an olive oil aerosol whose mean diameter is around 0.3μm. The 35mm drum camera, at 90° to the laser sheet propagation direction, records the Mie scattered laser light. The laser is pulsed at 12.5kHz with a pulse duration of ~20ns, the drum camera rotates at 250ms^{-1} as a streak camera, the pulsing of the laser effectively produces static frames on the film. Fifty images are recorded in each experiment.

Figure 2 shows eight sample images from the 50 recorded in a single experiment. This is an experiment with a trapezoidal notch perturbation on the upstream gas boundary. The gases are air/SF$_6$/air, SF$_6$ (sulphur hexafluoride) is ~6 times denser than air. In the images the illumination is from the light scattered from the olive oil droplets in the SF$_6$. These show: the initial condition (0.0ms is taken as the shock arrival time at the first air/SF$_6$ boundary); at 0.5ms the incident shock is half way through the dense gas; at 1.3ms a mushroom shaped air pocket is pushing into the SF6; at 1.9ms the shock has been reflected from the end wall and is now half way back through the dense gas region. The reflection of the shock from the end wall has now stagnated the gas flow, however the dense gas has momentum which causes further mixing as the mushroom air void is squashed and the air between the SF6 and the end wall is compressed and slows and pushes back the dense gas.

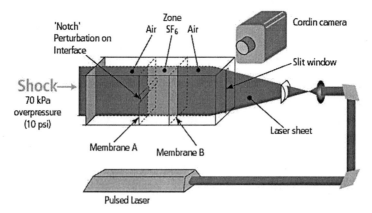

Figure 1. Illustration of laser sheet set-up.

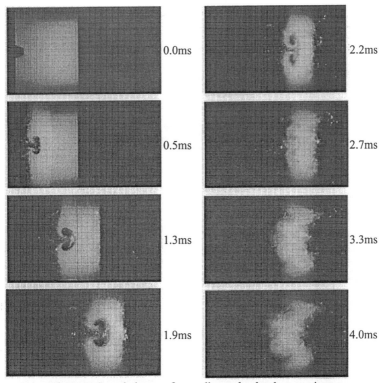

Figure 2. Sample images from a linear shock tube experiment

A video showing the capability of recording 50 images from a single experiment is available in the Hydrodynamics section within the Scientific and Technical - Featured Areas of the AWE web site (*www.awe.co.uk*). This and similar video sequences aid further understanding of the turbulent mixing in these experiments.

3. Shadowgraph imaging of detonation waves

The small detonation test-cell (SDT) gas chamber represents 1/10 of the CST annular gas chamber. It is constructed from Perspex (Lucite) slab sides with a 50mm separation, it is 54mm wide allowing placement of three miniature sparkplugs in the base with the same separation as in the CST. There is a half space separation between the two outer sparkplugs and the walls. It has two sections as shown in the photograph in figure 3, the lower is the detonable gas chamber which is 60mm high with a 60mm high expansion section above. An aluminium foil, thickness 17μm, is sandwiched between the two sections forming a gas tight seal. Ignition of the detonable gas by the sparkplugs causes detonation waves to propagate out from the sparkplugs which can be seen to coalesce near to the aluminium foil. The resultant overpressure caused by the detonation breaks the foil, which travels into the expansion section pushing air ahead of it. This forms a shock travelling ahead of the foil.

Figure 3. Detonation test cell.

A copper vapour laser pulsed at around 19kHz provides illumination for a shadowgraph system. A pair of matched 13-inch diameter mirrors produces parallel light through the test cell. A 70mm drum camera rotating at 300ms^{-1} captures the images. Again the pulsed laser produces individual frames on the film. With image separations at 50µs (minimum), longer than the transit time (23µs) of the detonation wave from spark ignition to the foil, it was necessary to perform several experiments to build up the sequence of images shown in figure 4. These show the detonation waves produced from three simultaneous sparks propagating into the detonable gas and combining together.

The images in figure 5 are subsequent images, in the upper expansion section, from those in figure 4 and show the air shock leading the foil, which has been accelerated by the detonation products. A gas mixture of 60/40 oxygen/acetylene (fuel rich) produced mean velocities (Figure 6) of: detonation velocity 2625 ms^{-1}; air shock velocity 1055ms^{-1}; foil velocity 825ms^{-1}. The air shock overpressure within the extension is approximately 11 bar.

4µs

19µs

Figure 4. Detonation wave propagation from 3 simultaneous sparks (shadowgraphy)

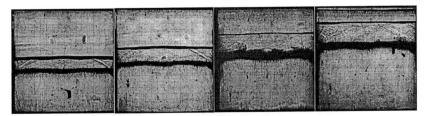

Figure 5. Air shock followed by aluminium foil

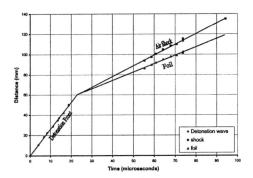

Figure 6. Detonation, foil and air shock velocities

4. Convergent Shock Tube

The use of the SDT has allowed improvement to be trialled before being implemented into the CST. Recent incorporation of these improvements into the CST has produced a working system. Imperfections evident in the shock as viewed in the CST apex region are believed to result primarily from known irregularities in the joints between the various sections. An improved version is currently being designed. Figure 7 illustrates the CST design. A sequence of selected shadowgraph images from a single convergent mix experiment is shown in figure 8. These show the incident shock moving up through the dense gas region (at the bottom) and then travelling through the air above before being reflected from the apex. Here the laser is pulsed at 15kHz.

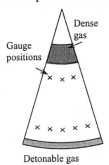

Figure 7. Illustration of CST design

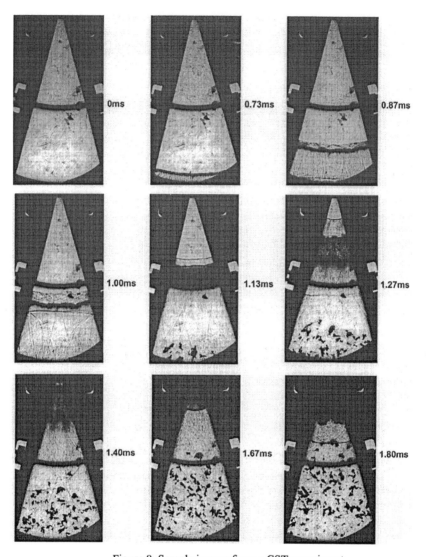

Figure 8. Sample images from a CST experiment

5. Proposal for New Imaging Technique

The linear shock tube at AWE has for a number of years produced some very good quality images and data from laser sheet images of Richtmyer-Meshkov turbulent mixing experiments. Recently the development of the Convergent Shock Tube (CST) has progressed and it has been used to conduct mix experiments. During the development, and for the mix experiments, shadowgraphy has been used as the primary diagnostic.

Laser sheet imaging requires the use of a seeding agent within one of the gases to scatter light into a camera. This therefore has the limitation of not being able to image the

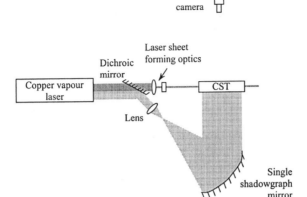

Figure 9. Illustration of plan view of proposed diagnostic

shock passage except where it passes through the seeded gas region. This poses no problem in the linear shock tube experiments, as the shock is stable when travelling within the confines of a rigid walled constant cross-section tube.

Shadowgraph imaging records the second differential of refractive index. And therefore records shock position well. It does not, however, differentiate between different gases very well. A significant problem in the CST experiments is that as the dense gas region is compressed and consequently compresses the air above it to a pressure of around 40 bar, no light is recorded on the film. Therefore the mixing between these gases is not recorded. Additionally there is no means of extracting density data to be able to identify different mix fractions.

A new imaging technique that might combine the benefits of both these two techniques is presented for discussion.

A copper vapour laser produces two wavelengths of light, 511 and 578nm, green and yellow, in an approximate energy ratio of 2:1. Laser sheet requires more pulse energy as the Mie scattering is very inefficient. Shadowgraphy with a laser requires the use of neutral density filters to reduce the light intensity before the camera. It is therefore proposed to use the green light for laser sheet and the yellow light for shadowgraphy. The two wavelengths can be easily split using a dichroic mirror as illustrated in figure 9.

Use of a colour film would allow the recording of both wavelengths on the same film. Printing with a single wavelength illumination or using band pass filters with a spectral source would, presumably produce prints of only one technique. Hence prints of two techniques from one film.

6. Film and ICCD camera discussion

As mentioned previously streak camera and a pulsed copper vapour laser are used to record 50 images per experiment. The use of photographic film does, however, have certain drawbacks.

Analysis of the experimental images is performed using computers and image analysis software. This requires the film to be processed and digitised, which is both costly and time consuming for the high resolution images required. Also each batch of photographic film has a non-linear characteristic curve which must be taken into account when analysing the images.

An intensified CCD camera has recently been procured with the intention of replacing the film camera in the shock tube imaging system. This camera is a 1300 x 1030 interline CCD camera with anti-blooming and is capable of 12 frames per second at maximum resolution of 12-bit accuracy, better than the 8-bit images previously obtained from digitising photographic images. The CCD has two stage Peltier cooling with secondary forced air cooling greatly reducing dark noise. The intensifier can be gated thus acting as an electro-optical shutter allowing nanosecond exposure times enabling the capture of a single laser pulse.

CCD cameras have many advantages over photographic film primarily that they produce a digital image negating the need for film processing and digitisation and the associated costs. This also greatly reduces the time taken to process and analyse an image from an experiment. However the main reason for purchasing an ICCD camera is its linear response thus making analysis easier and allowing quantitative data extraction, however this capability is limited at present due to multiple scattering as mentioned in our other poster presentation.

A significant drawback of current CCD cameras is the low frame rates that are achievable due to the readout time of the CCD chip. The camera purchased by AWE has a maximum frame rate of 12 frames per second therefore only one image can be captured during the 4ms recording time per experiment. There is however a possibility of implementing an array of four (or more) cameras to give multiple images of an experiment.

At present using one ICCD camera alongside the streak camera enables a high quality single image to be analysed and a video to be made using the 50 images allowing better visualisation of the mixing processes imaged.

7. Conclusions

The work presented here is a small sample of the work performed using the AWE shock tubes. This paper illustrates the uses to which our pulsed lasers and drum cameras are put in recording shocked and detonated gases and illustrates the advantages of obtaining a large number of high quality images. Results from both the laser sheet technique and shadowgraphy have been shown and are considered to be most effective for our applications. Some restrictions, particularly with obtaining quantitative data from these have been found and are reported in a separate poster. An idea for a merging of the two techniques has been discussed and will be attempted at a later date. The relative benefits of CCD cameras over film have been briefly discussed.

8. References

[1]   Landeg D, Philpott M, Smith I and Smith A. 1993. *The Laser Sheet as a Quantitative Diagnostic in Shock Tube Experiments.* Proceedings of the 4th IWPCTM – Cambridge, England.

[2]   Cowperthwaite N, Philpott M, Smith A V, Smith I D and Youngs D. 1995. *Shock-Induced Instability Experiments on Gas Interfaces Featuring a Single Discrete Perturbation.* Proceedings of the 5th IWPCTM – Stony Brook, USA.

[3]   Smith A V, Philpott M K, Millar D B, Holder D A, Cowperthwaite N W and Youngs D L
        1997. *Shock Tube Investigations of the Richtmyer-Meshkov Instability due to a Single Discrete
        Perturbation on a Plane Gas Discontinuity.* Proceedings of the 6th IWPCTM – Marseille,
        France.

[4]   Smith A V, Holder D A, Philpott M K and Millar D B. 1999. Notch and Double Bump
        Experiments using the 200 x 100mm Linear Shock Tube. Proceedings of the 7th IWPCTM –
        St. Petersburg, Russia.

Inst. Phys. Conf. Ser. No. 177
Paper presented at 1st Int. Conf. on Optical & Laser Diagnostics, London, 16–20 Dec. 2002
©2003 IOP Publishing Ltd

An Investigation of the Problem of Multiple Scattering from Seeding Particles in Laser Sheet Studies of Shock Induced Gas Mixing

David A Holder, Martin K Philpott and David L Youngs

AWE, Aldermaston, Reading, Berkshire, RG7 4PR, UK

Abstract. Analysis of shock tube turbulent mixing experiments, using a pulsed laser sheet with Mie scattering from a seeded gas, has highlighted a complex data extraction problem. Its source is high-order multiple scattering from the seeding particles. The aim of the experiments is to validate a gas mixing computational code. Meaningful quantitative data from the experimental images is essential for code validation, yet seemingly impossible to extract. Therefore the code results were modified to include representation of multiple scattering.

This paper outlines the initial attempt to analytically apply Mie scattering theory to describe the observed multiple scattering. Limited to primary and secondary scattering, it inadequately describes the experimental observations. A successive method employed a Monte Carlo technique simulating up to 10th order scattering superimposed on the computational results. Details of this method are presented. Its success is demonstrated by comparison of experimental images with the modified code results and line-out plots of scattered intensity.

Longer-term solutions to eliminate multiple scattering are discussed, including results from an Intensified CCD camera which enables a reduced seeding concentration and hence less multiple scattering. It shows the laser sheet technique capable of directly producing quantitative mix data.

1. Introduction

A programme of shock tube turbulent mixing experiments using a pulsed laser sheet with Mie scattering from a seeded gas has highlighted a complex data extraction problem due to image blurring. The source of the problem has been identified as the presence of high-order multiple scattering from the seeding particles. The aim of the experiments is to investigate shock induced dynamic mixing of air with a dense gas (SF_6, sulphur hexafluoride) and provide quantitative data to validate a Richtmyer-Meshkov turbulent gas mixing code. Validation is by comparison between the experimental images and their corresponding plane 'slice' through a full 3D numerical simulation. The simulation is by the AWE TURMOIL 3D code which is a semi-Lagrangian calculation in which the mesh moves with mean fluid velocity. Use of a relatively high seeding concentration is necessarily dictated by the sensitivity of the photographic film and available laser pulse energy. Meaningful quantitative data on gas mixing is essential for code comparison, yet seemingly impossible to extract from the recorded images. Code validation then becomes dependent on qualitative comparison of the experimental blurred images with the modified code 'images'.

An initial attempt to analytically apply Mie scattering theory to describe the observed multiple scattering process, limited to primary and secondary scattering, was shown to be incapable of adequately describing the observed dependence of the scattered light intensity on propagation distance. Therefore, a successive method employed a Monte Carlo technique, simulating up to 10th order scattering, superimposed on the gas mixing code results. The paper focuses on this Monte Carlo post-processing technique with graphical presentation of the relative contribution from each of the scattering orders. Success is demonstrated by the close comparison of the experimental images to those derived from the modified code; similarly with sample line-out plots of scattered intensity as a function of distance.

2. Experiment (Application)

The schematic in figure 1 illustrates the experimental application of this work. It shows the test cell of the shock tube with a plane (A) and a profiled (B) microfilm membrane constraining a dense gas, which is seeded with an olive oil aerosol. The test cell is 200mm high and 100mm wide, the dense gas region is 150mm long. This is illuminated with a pulsed copper vapour laser sheet, 200mm high and 2mm wide. The Mie scattered light is recorded with a Cordin drum camera viewing at 90°. A flat-topped shock of 70kPa overpressure and 7ms duration passes through the dense gas region fragmenting the membranes and initiating Richtmyer-Meshkov Instability mixing.

3. Example of Multiple Scattering in a Laser Sheet Image

The images in figure 2 show the 'blurring' effect of multiple scattering. The image on the left is a direct print from the photographic negative, the image on the right is a central slice through the full three dimensional code calculation of the experiment showing partial gas density: both represent the same time, 4.0ms. When examined closely, the large scale features of each are the same. The experimental image suffers from a lack of laser intensity at the top and bottom, but otherwise the two are comparable.

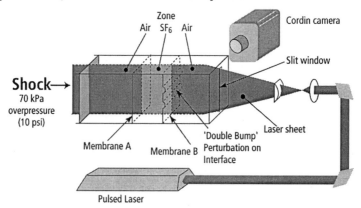

Figure 1. Illustration of laser sheet set-up.

Experiment   Code

Figure 2. Images showing the effects of multiple scattering

The major difference between code and experiment is that the experimental image shows a region of almost uniform light intensity, whereas the code result shows a significant amount of fine structure. The lack of this structure in the experimental image is due to the presence of multiple scattering of the light within the seeded gas region.

4. Light Scattering

The illustration and graph to the left of figure 3 show primary scattering and the expected decay of the scattered light intensity as a function of distance for a uniformly seeded gas region. The graph to the right shows the type of result actually found by experiment and above it an illustration of how higher orders of scattering change the expected intensity.

5. Monte Carlo Simulation

The Monte Carlo code simulation of multiple scattering uses Mie theory for the mean scattering cross section. It then assumes a known droplet size distribution and concentration. An input of 800,000 particles is modelled to represent a laser sheet pulse. The simulation models up to 10th order multiple scattering. Only the first scattering event is necessarily within the plane of the laser sheet. The SF_6 gas distribution calculated by the 3D numerical turbulent mixing simulation is input into the Monte Carlo post-processor to produce a scattered image.

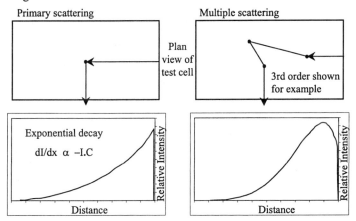

Figure 3. Graphical description of primary and multiple scattering

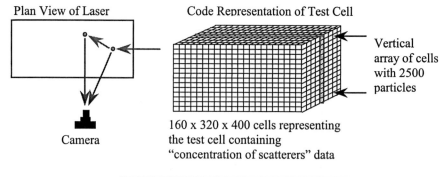

Plan View of Laser

Code Representation of Test Cell

Vertical array of cells with 2500 particles

Camera

160 x 320 x 400 cells representing the test cell containing "concentration of scatterers" data

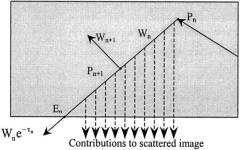

Figure 4. Illustration of the Monte Carlo simulation.

A computational particle represents a packet of photons of energy W. Each particle is forced to have n (~ 10) collisions and the scattered light intensity contribution to the image for each order of collision is calculated.

W_n = Particle energy after collision n at point P_n

τ_n = Number of mean free paths along exit path P_nE_n

$W_n e^{-\tau_n}$ = energy lost from the system

$W_{n+1} = W_n\left(1 - e^{-\tau_n}\right)$, energy after n+1 collisions at a randomly chosen point P_{n+1}

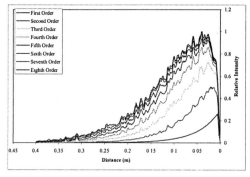

Figure 5. Graph showing the effects of multiple scattering contributions

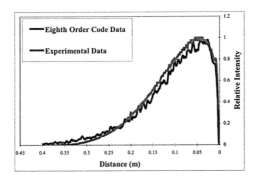

Figure 6. Comparison between experimental and scatter corrected code lineouts

6. Success of Monte Carlo Simulation

The graph in figure 5 shows the effect of the different scattering 'orders' on the amount of light reaching a camera as predicted by the Monte Carlo post-processing code for a uniformly seeded gas region. Figure 6 illustrates the success of the Monte Carlo code correction in matching the experimental data for a uniformly seeded gas region. This success is reflected in the relatively close match between the experimental and code + scattering images shown in figure 7. Also included are images that are slices through the full 3D numerical simulation. To further demonstrate the success of the Monte Carlo post processor developed at AWE simple lineouts are presented comparing both experimental and code + scattering images. Figure 8 shows the comparison for two times, 1.9ms and 2.7ms.

7. Conclusions

Post-processing using the Monte Carlo simulation of Multiple Scattering provides a useful comparison between code and experimental images. However, this is not an ideal situation as there are two steps involved which could compromise code validation. An experimental solution is required.

A significant reduction in the seeding concentration to increase the scattering mean free path would render the multiple scattering component negligible but would also reduce the light intensity at the camera. Conventional film cannot cope with the reduction required (typically a factor of ~10). It is therefore the intention to use an Intensified CCD camera which will allow recording of the low light levels.

A reduction in the light intensity will, however, cause a corresponding increase in the relative intensity of reflections from the membrane fragments and test cell components. A proposed solution is the use of PLIF (Planar Laser Induced Fluorescence), which emits at longer wavelength, and a corresponding long pass filter on the camera.

A single frame ICCD camera has been purchased and is undergoing trials. It is anticipated that this will be incorporated into a suite of such cameras to provide a time history of the experiments.

236

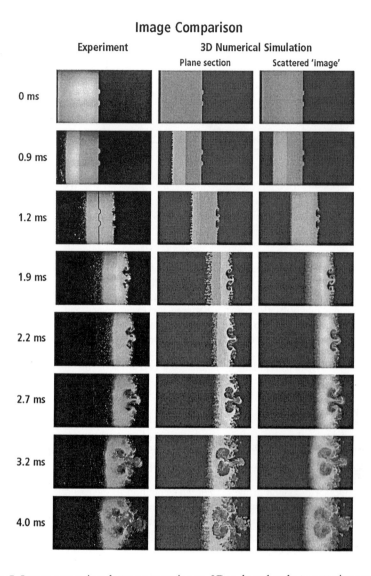

Figure 7. Image comparison between experiment, 3D code and code + scattering correction.

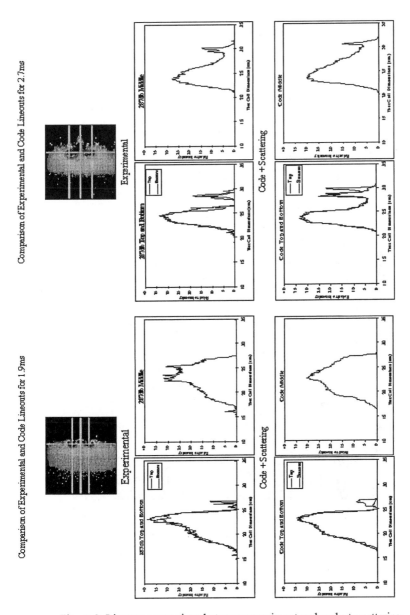

Figure 8. Lineout comparison between experiment and code + scattering.

Inst. Phys. Conf. Ser. No. 177
Paper presented at 1st Int. Conf. on Optical & Laser Diagnostics, London, 16–20 Dec. 2002

239

PIV Measurements in Large-scale Cardiovascular Models – The Importance of Dynamic Scaling

I Owen, J D Gray and M P Escudier

Department of Engineering
University of Liverpool
Liverpool, L69 3GH
United Kingdom

Abstract. Particle Imaging Velocimetry and Laser Doppler Anemometry are often used in research into blood flow (haemodynamics). This paper shows how a large-scale model of an arterial junction (a distal anastomosis, or bypass graft) can be used to represent the in vivo blood flow provided the correct scaling procedures are applied. Blood is a non-Newtonian liquid whose viscosity can be represented by a power law. Dimensional analysis shows that a pulsatile flow of a power-law fluid can be characterised by a generalised Reynolds number, the Strouhal number, the Pulsatility Index, and by the power-law exponent n. It is clearly demonstrated in this paper that a non-Newtonian liquid cannot be properly represented by a Newtonian liquid, but it can be very well represented by a non-Newtonian liquid that has the same power-law exponent. If, however, the common practice of using a Newtonian liquid is adopted, the wall shear stress along the bed of the anastomosis will be seriously over-predicted by as much as 100-200%.

1. Introduction

Knowledge of the fluid flow within the cardiovascular system is important for the understanding of atherosclerosis, a disease of the medium to large arteries involving the build-up of fatty deposits, and the biggest killer in the western world. Arteries are typically 4–6 mm in diameter, and are complex in their construction, shape and behaviour. In recent years LDA and PIV techniques have been applied to this field of research in an attempt to understand the detailed nature of the flow. Since it is extremely difficult to make detailed flow measurements *in vivo*, physical models that accurately represent the flow conditions in the cardiovascular system are needed. The great majority of measurements are made in life-scale geometries in the belief that this is the best way to reproduce the correct flow conditions. Also, blood is known to be a non-Newtonian liquid, but the majority of studies employ Newtonian liquids in the belief that the non-Newtonian behaviour does not have an important effect. A major weakness of working with small-scale models is that it makes it more difficult to obtain good resolution in the measurements. An engineering approach to this problem would be simply to increase the size of the model, identify the appropriate dimensionless groups, and apply the necessary scaling procedures. This approach does not appear to have been adopted in mainstream haemodynamics research.

Scaling methods for Newtonian fluid flow are well established and are in widespread use throughout engineering research and practice. In the case of blood, which is non-Newtonian, the scaling is more complex. One reason for the scaling complexity is that it is difficult to separate the effects of Reynolds number from the viscous (or rheological) effects in any comparison between the flow of fluids of a different rheological character. Various methods for calculating 'equivalent' flow rates for blood flow are currently employed in the literature, but there is uncertainty as to which is most appropriate. For this reason, a suitable approach to scaling is needed for experiments using both non-Newtonian and Newtonian blood 'analogues'.

The specific case under consideration in this study is the flow in an arterial bypass graft (a distal anastomosis). Ideally, any model used to study this (or any other) type of flow will be dynamically similar to the *in vivo* case, i.e. all the relevant forces will be in the same ratio for both model and life scale and therefore all the flow phenomena at life scale will be reproduced in the model. Since blood is a non-Newtonian fluid, this would clearly involve including at least one parameter to describe its non-Newtonian characteristics. However, in an effort to simplify what is a complex collection of conditions, many authors [e.g. 1, 2, 3] have chosen to neglect the non-Newtonian character of blood. In the case of blood flow through an artery, for example, the flow of blood *in vivo* is often modelled using a Newtonian liquid *in vitro*. This has generated considerable debate as to how big an effect this has on the flow conditions being studied.

Unfortunately, despite many years of research there is still considerable confusion in the literature about the comparison between non-Newtonian and Newtonian blood analogues. Some authors [e.g. 4] report no differences in the essential flow characteristics, with errors limited to 50% in the prediction of secondary flows. The majority of authors conclude that there are differences, but due to the myriad of different flow conditions in the body, cannot come to a consensus about the magnitude of these differences, or the conditions under which they appear. For example, Ballyk *et al* [5] report in a study of a distal anastomosis that the rheological effects are important in steady flow conditions, but largely disappear under pulsatile conditions, while Rodkiewicz *et al* [6] conclude that for a straight artery, the non-Newtonian effects only appear in pulsatile flows. Chakravarty and Mandal [7] also find seemingly contradictory results, reporting that a Newtonian fluid can model blood flow satisfactorily in the parent aorta of the aortic arch, but 'drastic change' is seen further downstream in the daughter artery.

Much of the confusion arises from the fact that most authors do not give due attention to the scaling procedures that arise from a process of dimensional analysis. The approach taken in much of the literature is to assume the non-Newtonian fluid (blood) can be satisfactorily represented by a Newtonian fluid if the viscosity is chosen correctly. Doing this makes it possible to then study the two fluids (of different rheological character) at apparently similar Reynolds numbers. In general there are two schools of thought for the best method of generating a representative viscosity for blood. Since blood has a high shear Newtonian plateau, many [e.g. 8] use this value as a representative viscosity, e.g. 3.5 mPa.s. Others [e.g. 9] prefer to take into account more of the low shear behaviour of blood and choose to use a higher 'characteristic' viscosity, e.g. 7.8 mPa.s.

2. Dimensional Analysis and Scaling

In this study it is assumed that blood rheology can be described by a power law relationship where the apparent viscosity, η, is a function of the consistency index, κ, the shear rate in the liquid, $\dot{\gamma}$, and the power-law index, n, as shown in eqn.1. It is further assumed that the blood flows with (spatial) mean velocity U, through rigid-walled arteries of diameter D.

$$\eta = \kappa \, \dot{\gamma}^{(n-1)} \tag{1}$$

The wall shear stress, τ_w, is considered to be an important parameter since areas of low shear are identified with areas prone to atherosclerosis. For steady flow conditions, τ_w can be seen to be a function of:

$$\tau_w = f\,(U, D, \rho, \kappa, n) \tag{2}$$

where ρ is the liquid density. This leads to the general relation:

$$\frac{\tau_w}{\tfrac{1}{2}\rho U^2} = f\left[\frac{\rho D^n \, U^{(2-n)}}{\kappa}, n\right] \tag{3}$$

The first term in the bracket can be regarded as a generalised Reynolds number, Re(G). The inclusion of n as a separate dimensionless 'group' is expected since it is a dimensionless parameter. However, its presence in the general relation emphasises the fact that two fluids cannot be dynamically similar if the exponent n is not the same, i.e. a non-Newton liquid cannot be modelled by a Newtonian one. If n = 1, i.e. the liquid is Newtonian, eqn.(3) takes on the usual form of friction coefficient being a function of Reynolds number.

An additional characteristic of blood flow is that it is pulsatile in nature, so in addition to the parameters included in eqn.(2), the flow will also be characterised by the pulsation frequency, ω, and the range of velocity, $\hat{u}$ (= $u_{max} - u_{min}$). Thus:

$$\tau_w = f\,(U, D, \rho, \kappa, n, \omega, \hat{u}) \tag{4}$$

where U is now both the spatial and temporal mean velocity. This leads to the general relation for pulsatile flow:

$$\frac{\tau_w}{\tfrac{1}{2}\rho U^2} = f\left[\frac{\rho D^n \, U^{(2-n)}}{\kappa}, n, \frac{\omega D}{U}, \frac{\hat{u}}{U}\right] \tag{5}$$

$\dfrac{\omega D}{U}$ is the Strouhal number, and $\dfrac{\hat{u}}{U}$ the pulsatility index.

Experiments were therefore designed in which these dimensionless groups were used to maintain dynamic similarity between measurements involving Newtonian and non-Newtonian liquids. Figure 1 shows a collection of blood data taken from numerous sources (see [10]). The best fit to this data produces a power-law index n = 0.63 and a Consistency Index κ = 15 mPa.s. Therefore, to represent blood a non-Newtonian liquid was produced using an aqueous solution of 0.07%wt Xanthan Gum (a high molecular weight polymer). Its rheological characteristics were measured using a TA *AR1000N* cone and plate rheometer; n was found to be 0.63, κ was found to be 35.6mPa.s, and its density was found to be 1002 kg/m^3. An aqueous solution of 60%wt glycerine (viscosity = 11.0 mPa.s, density = 1152 kg/m^3) was used as the Newtonian fluid. Also shown on Fig. 1 are three Newtonian viscosities including the high-shear plateau of 3.5 mPa.s and the 'characteristic' viscosity of 7.8 mPa.s. To provide a greater range of viscosities to compare with the non-Newtonian case the value of 2.0 mPa.s was also considered.

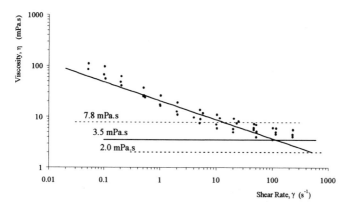

Figure 1 Non-Newtonian characteristics of blood

3. Experimental Methods and Procedure

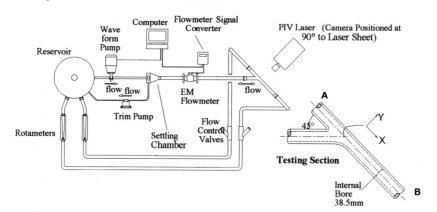

Figure 2 Schematic diagram of the experimental apparatus

The apparatus is shown schematically in Fig. 2. The pipework was constructed primarily from u-PVC plastic, while borosilicate glass was used to manufacture the testing section. This comprises a simplified model of a distal anastomosis (bypass graft) with a bore diameter of 38.5mm, together with a straight entry length of 47 pipe diameters.

The flow rig can be used to produce any shape waveform that has been pre-programmed into the computer, which in turn sends a variable voltage signal to the 3-phase motor of the main pump. Since the flow rate from a gear pump is directly related to the speed of the pump, it is relatively straightforward to set up the desired waveform in the testing section. However, the main pump operates only in the forward direction and ideally should always operate above zero flow to avoid any discontinuities in the required waveform. To overcome this problem a second gear pump was included to pump fluid at a steady rate out of the settling chamber straight back into the reservoir. This has the effect of shifting the entire waveform downwards, and allows waveforms that have reverse portions to be generated in the testing section. The pulsatile flow rate was measured by the electromagnetic flowmeter which had been specially designed by Fischer & Porter to measure unsteady flows. Figure 3 shows the waveform that was used. The solid line represents the profile supplied to the pumps, and the symbols represent the flow rates measured by the flowmeter in two consecutive cycles. The agreement is very good and repeatable. The phase shift between the input and output signals simply reflects the time required for the fluid at the flowmeter to respond.

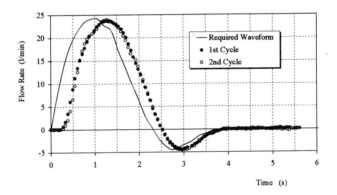

Figure 3 Biphasic waveform used in pulsatile tests

Once a relevant waveform flow had been set up in the rig, the flow through the testing section could be studied. This simplified distal anastomosis consisted of a 45° junction manufactured from borosilicate glass tubing. The junction represents the downstream (distal) junction (anastomosis) of a bypass graft. The main flow into the junction is from the graft and A-B (Fig. 2) represents the original artery that will have a blockage (occlusion) in the proximal branch. In many situations, when the flow passes through a surgical bypass of this kind, the bulk of the exit flow will be in the distal branch (downstream, B) but there may be some proportion of the flow that exits the anastomosis from the proximal branch (upstream, A). Two flow control valves were positioned downstream of each branch of the testing section to allow the flow to be split in any distal:proximal ratio. Each of the three ratios used in the study (100% distal, 75% distal, and 50% distal) were measured, under steady flow conditions, by the Rotameters located at the end of each return branch. The whole anastomosis model was housed in a refraction box filled with pure glycerine which has the same refractive index as the borosilicate glass. On the inside surface of the testing section it was not possible to obtain a perfect refraction match since the properties of the blood analogues were chosen to obtain the correct rheology.

The Dantec PIV System comprised a pulsed Nd:YAG laser (NewWave Research Minilase III, 532nm wavelength), a digital video camera (Dantec DoubleImage 700) and a purpose built processor that forms the core of the vector processing unit. This hardware was used in conjunction with the FlowManager software. Because of the curvature of the glass tubes, and to allow for any refractive index mismatch, a calibration target was inserted into the centre-line of the pipe prior to experimentation and used to determine the true position within the flow field. The fluid was seeded with 75μm diameter fluorescent particles. For steady flow measurements the PIV system was activated at will and 100 samples were taken and averaged. For the pulsatile tests, a point in the cycle was selected and the external trigger of the PIV system was synchronised with the signal to the pump so that 100 samples were taken at identical points in the cycle, and then averaged (they were also individually compared to ensure their consistency). The laser sheet passed through the central plane of the glass model, as indicated in Fig. 2.

The experiments (steady and pulsatile) were conducted in two separate parts using the non-Newtonian and Newtonian liquids. The experimental rig is approximately six times bigger than a life-scale artery (the femoral artery typically has a diameter of 6 mm). For the power-law fluid the scaling procedure outlined by eqn.(5) was employed, while for the Newtonian case a suitable Reynolds number range (Re(N)) was found to be 130 – 1200. The *in vivo* flow rates are typical of conditions in the femoral artery. Table 1 summarises these values and their large-scale equivalents.

Table 1   Generalised Reynolds Numbers
calculated from *in vivo* parameters

Flow rate *in vivo* (l/min)	Mean velocity (mm/s) *in vivo* diameter = 6mm	Re(G)	Flow rate in rig (l/min) (from scaling)
0.120	70.1	76	4.16
0.270	159.2	232	9.39
0.415	244.6	420	14.49
0.563	331.9	640	19.70

Results from different experiments were compared by considering the dimensionless shear stress at the wall. Figure 4 shows a typical vector map from the PIV. Thus the wall shear stress distributions along the bed of the junction (A-B in Fig. 2) were determined using the local velocities nearest to the wall, the distance of these velocities from the wall, and the rheological characteristics of the fluid. Estimates of axial wall shear stress were found using the following equation:

$$\tau_w = \kappa \left(\frac{u}{\delta} \right)^n \qquad (6)$$

where u is the velocity a distance δ from the wall (typically ≈ 1mm).

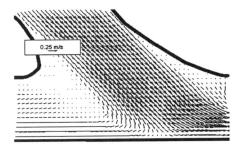

Figure 4 Velocity vector map for 0.07% Xanthan Gum, 75% distal split

4. Discussion of Results

The first test was to validate the scaling procedure by comparing the wall shear stress distributions for two non-Newtonian liquids. A second, thicker power law fluid (0.05%wt Xanthan gum, 50%wt Glycerine, aqueous) was used to obtain results that could be compared to the Xanthan gum results. The power law index of this fluid was the same as the 0.07%wt Xanthan gum solution, but the value of κ was increased (81.0 mPa.s *cf* 35.6 mPa.s). Dynamic similarity was obtained by adjusting the flow rates and cycle frequency to achieve similarity of the generalised Reynolds number, the Strouhal number, and the pulsatility index. A selection of results taken from the pulsatile tests can be seen in Fig. 5; similar data was also obtained from the steady flow experiments. The results show the wall shear stress measured for the two fluids at two different positions in the flow cycle, the distance origin being at the intersection of the centrelines of the two pipes.

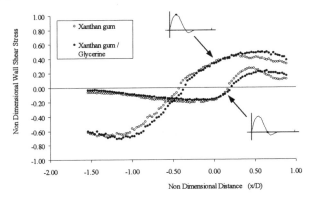

Figure 5 Wall shear stress along bed of artery for two different power law fluids demonstrating validity of scaling procedure. Re(G)=70, 50% distal split.

The general form of these graphs is seen in all the wall shear stress measurements. This particular figure shows the wall shear stress for two different positions in the flow cycle: the maximum flow (peak-prograde) and the point where the flow changes from positive to negative and is zero (end-prograde). Although the net flow rate is zero at this point in the cycle, since the flow is a dynamic one, features such as recirculations and secondary flows will still be present. The data in Fig. 5 is for a 50% distal split (i.e. 50% of the flow passes downstream via B, and 50% via A). Considering the maximum flow data, the flow entering the proximal branch represents a negative velocity and shear stress and the flow through the distal branch is considered positive. The point where the shear stress is zero represents the stagnation point. As can be seen the stagnation point changes location during the cycle.

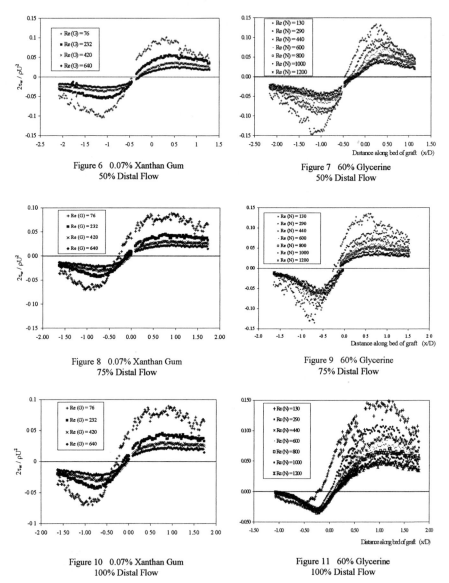

Figure 6   0.07% Xanthan Gum
50% Distal Flow

Figure 7   60% Glycerine
50% Distal Flow

Figure 8   0.07% Xanthan Gum
75% Distal Flow

Figure 9   60% Glycerine
75% Distal Flow

Figure 10   0.07% Xanthan Gum
100% Distal Flow

Figure 11   60% Glycerine
100% Distal Flow

The results shown in Fig. 5 were taken at very different flow rates and cycle times to obtain the Reynolds and Strouhal number equivalencies. The good agreement of the data demonstrates that scaling by dynamic similarity works with non-Newtonian liquids (as it does of course with Newtonian liquids). The question that remains to be addressed is whether the flow of a non-Newtonian liquid can be represented by that of a Newtonian liquid. Figures 6-11 show the results obtained for the 0.07%wt Xanthan gum solution and the Newtonian fluid, at the three different flow splits. The graphs are presented in a similar way to Fig. 5 and are for steady flow conditions. The general form of the shear stress distributions are the same for the non-Newtonian and the Newtonian liquids. However, for any pair of figures (i.e. same percentage distal flow), the flow characteristics for similar Reynolds numbers (generalised, Re(G) and conventional, Re(N)) are not the same. If one seeks the Reynolds numbers that do give similar shear distributions for one pair of figures, the same Reynolds numbers do not give comparable shear stress distributions for the other two pairs. Furthermore, if two flow rates, or Reynolds numbers, were established that gave similar shear stress distributions then even though the wall stresses may be similar, the flow structure (recirculations, secondary flows, etc) would not be the same. These comments are made on the basis of the steady flows shown in Figs. 6-11, when a pulsatile flow is used the mis-match between the two types of fluids will change through the flow cycle. In other words, a Newtonian liquid cannot be used to represent a non-Newtonian liquid since there is no way of achieving dynamic similarity.

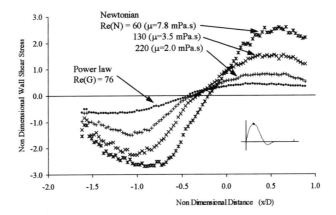

Figure 12 Comparison of wall shear stress for Newtonian and non-Newtonian liquids at same flow rates, 50% distal split.

It is common in haemodynamics research to work with life scale models and to use the same flow rates as those found *in vivo*, whatever the experimental fluid. Clearly, doing this takes no account of scaling by the dimensionless parameters and is simply scaling by equivalence of size and of fluid velocity. The invalidity of this approach is demonstrated in Fig. 12 which shows the results of an experiment that was carried out under pulsatile conditions. The figure is presented as the predicted stress that would be generated *in vivo*, using the power law-fluid to represent blood and using three Newtonian blood analogues having the viscosities presented earlier in Fig. 1. The Reynolds numbers were set (in the experimental rig) such that the equivalent flow rates of the *in vivo* case considered here would be the same for each of the four fluids. It can be seen that the shear stresses obtained by using equal flow rates leads to a very large over-statement of the wall shear stress. It will also lead to very different flow structures in the junction. This is important, since both wall shear stress and flow structure are known to have an effect on atherosclerosis.

5. Conclusions

PIV and LDA techniques are often applied to research into blood flow in the cardiovascular system. Arterial models are commonly used in experiments because it is difficult to make velocity measurements *in vivo*. To obtain greater resolution it is better to use large-scale models, and it is necessary to use a suitable liquid to represent the blood. While this approach is perfectly valid, the importance of using the correct dynamic scaling must not be overlooked. This paper has examined the common approach of assuming the blood can be represented by a Newtonian analogue. It has been found that by using the most popular Newtonian assumption (i.e. blood viscosity = 3.5 mPa.s), very large errors are generated in the prediction of the wall shear stress values along the bed of a distal anastomosis. Furthermore, if a higher value for blood viscosity is assumed, these errors will increase, not decrease. It has also been found that there is no single value of Newtonian viscosity that can be assumed to give a close numerical match with the non-Newtonian flow for every case. However, it has also been demonstrated that if blood is assumed to behave as a power-law fluid, then by using the correct scaling procedures a non-Newtonian fluid having the same power-law exponent as blood can confidently be used to study the flow behaviour of blood in large-scale arterial models.

References

[1]   Perktold K, Hofer M, Rappitsch G, Loew M, Kuban B D and Friedman M H 1998 *J. Biomechanics* 31 217-28
[2]   Ojha M, Cobbold R S C and Johnston K W 1994 *J Vascular Surgery* 19 1067-73
[3]   Hughes P E, How T V 1996 *J. Biomechanics* 29 855-72
[4]   Perktold K, Resch M and Florian H 1991 *J. Biomechanical Engg.* 133 464-75
[5]   Ballyk P D, Steinman D A and Ethier C R 1994 *Biorheology* 31 565-86
[6]   Rodkiewicz C M, Prawal Sinha and Kennedy J S 1990 *J. Biomechanical Engg* 112 198-206
[7]   Chakravarty S and Mandal P K 1997 *Int. J. Engng. Sci.* 35 409-22
[8]   Cho Y I and Kensey K R 1991 *Biorheology* 28 241-62
[9]   Gijsen F J H, Allanic E, van de Vosse F N and Janseen J D 1999 *J Biomechanics* 32 705-13
[10] Gray J D 2002 *Non-Newtonian scaling of blood flow in a femoral bypass* PhD Thesis University of Liverpool

Inst. Phys. Conf. Ser. No. 177
Paper presented at 1st Int. Conf. on Optical & Laser Diagnostics, London, 16–20 Dec. 2002
©2003 IOP Publishing Ltd

249

Laser diagnostics of dynamic phenomena in biological objects: model measurements, numerical simulations, and medical applications

A V Priezzhev, K B Begun, A Y Tyurina, O E Fedorova, N N Firsov*,
S J Matcher**, and I V Meglinski***

Lomonosov Moscow State University, Moscow, Russia,
*Pirogov Russian State Medical University, Moscow, Russia
**Exeter University, Exeter, UK
***Cranfield University, UK

Abstract. Motion is an intrinsic feature of the live objects at all organization levels: molecular, cellular, tissue, organ, and whole organism. Laser diagnostic techniques based on elastic, quasielastic, and inelastic light scattering, fluorescence, interferometry, etc. have been widely applied to detect and quantitatively measure the parameters of biological motion.

The paper gives a brief overview of these applications and focuses on discussion of laser diagnostics of dynamic phenomena in blood and blood perfused tissues. In particular the application of laser Doppler microscopy and flowmetry, optical coherence tomography (OCT), and diffusing wave spectroscopy (DWS) are discussed in relation to the assessment of blood flow parameters in single vessels and in depth the skin. Measurements of red blood cell dynamic interactions resulting in cell membrane deformation, cell aggregation and orientation in shear flow, etc., by means of laser scattering are discussed with reference to clinical value and relation to different diseases. Different approaches that enable such laser measurements and are helpful for data evaluation are outlined. They include: application of cw and time-gating techniques with different scheme arrangements;- application of long- and short-coherence length interferometry; - Monte-Carlo simulation of the detected optical signal in different regimes.

Inst. Phys. Conf. Ser. No. 177
Paper presented at 1st Int. Conf. on Optical & Laser Diagnostics, London, 16–20 Dec. 2002
©*2003 IOP Publishing Ltd*

Study of Flow Characteristics and Particle Trajectories in a Model of the Human Upper Airway

H K Versteeg and G K Hargrave

Wolfson School of Mechanical and Manufacturing Engineering, Loughborough University, Loughborough, Leicestershire, UK.

Abstract. Air inhaled into the upper airway flows through a sequence of anatomical structures including: oral cavity and oropharynx; epiglottis; larynx; trachea and left and right main bronchi. From previous research, it is clear that the geometry of the upper airway has a major effect on the internal flow characteristics and deposition patterns of inhaled medical aerosols. Increasingly CFD is being used to understand these flows, but there is a paucity of validation data to aid development and validation. Ideally one could consider the relevant transport processes separately in the above anatomical structures, but this is not possible since the flow characteristics are known to be sensitively dependent on inlet flow conditions. The fluid engineering literature contains numerous examples of strong interactions, particularly in flows with large streamline curvatures such as those encountered in the current geometries. Recirculating flows and jet flows are also highly susceptible to details of downstream geometry and boundary conditions.

In this paper we present the findings of an in-vitro study of flow characteristics and aerosol particle behaviour in a section of the upper airway located just upstream from the larynx down to and including the bifurcation of the trachea into the main bronchi. We give results of steady flow tests under laminar and turbulent flow conditions using high-speed imaging and Particle Image Velocimetry (PIV) to characterise the flow and particle behaviour in our model. We report the main flow structures and highlight the interplay between geometry and flow. We also examine the relationship between particle trajectories and flow structures with particular emphasis on geometrical interaction effects.

Inst. Phys. Conf. Ser. No. 177
Paper presented at 1st Int. Conf. on Optical & Laser Diagnostics, London, 16–20 Dec. 2002
©2003 IOP Publishing Ltd

Glucose Monitoring using Laser NIR Spectroscopy

V A Saetchnikov[1,2], E A Tcherniavskaia[2]

[1]Ruhr-Universitaet, Bochum, Germany
[2]Belorussian State University, Minsk, Belarus

Abstract. Non-invasive method for control concentration of glucose based on NIR absorption and fluorescence tissue spectroscopy has been developed. Equipment includes laser, silica fiber, spectrograph with CCD camera and computer. Spectrometry data proceeding includes partial least squares (PSL) regression analysis combined with neural network algorithms used to evaluate glucose concentrations in blood and water from approximately 100 mg/dl and 50 mg/dl.

Method is combined with NIR Raman spectroscopy of glucose in water, acetone and blood to increase a reliability of obtained data and a sensitivity for minimal glucose concentration. Obtained results are used for development of a compact set of equipment for noninvasive NIR laser glucose monitoring.

Author Index